Lecture Notes in Bioinformatics 10477

Subseries of Lecture Notes in Computer Science

More information about this series at http://www.springer.com/series/5381

Andrea Bracciali · Giulio Caravagna
David Gilbert · Roberto Tagliaferri (Eds.)

Computational Intelligence Methods for Bioinformatics and Biostatistics

13th International Meeting, CIBB 2016
Stirling, UK, September 1–3, 2016
Revised Selected Papers

Springer

Editors
Andrea Bracciali 🅾
Computing Science and Mathematics
University of Stirling
Stirling
UK

Giulio Caravagna 🅾
School of Informatics
University of Edinburgh
Edinburgh
UK

David Gilbert 🅾
Department of Computer Science
Brunel University London
Uxbridge, Middlesex
UK

Roberto Tagliaferri 🅾
Department of Management and Innovation
  Systems DISA-MIS
University of Salerno
Fisciano
Italy

ISSN 0302-9743   ISSN 1611-3349 (electronic)
Lecture Notes in Bioinformatics
ISBN 978-3-319-67833-7   ISBN 978-3-319-67834-4   (eBook)
DOI 10.1007/978-3-319-67834-4

Library of Congress Control Number: 2017955205

LNCS Sublibrary: SL8 – Bioinformatics

Printed on acid-free paper

This Springer imprint is published by Springer Nature
The registered company is Springer International Publishing AG
The registered company address is: Gewerbestrasse 11, 6330 Cham, Switzerland

Preface

The 13th annual edition of the international meeting on Computational Intelligence methods for Bioinformatics and Biostatistics (CIBB 2016) built upon the tradition of the CIBB conference series and provided a multi-disciplinary forum open to researchers interested in the application of computational intelligence, in a broad sense, to open problems in bioinformatics, biostatistics, systems and synthetic biology, medical informatics, as well as computational approaches to life sciences in general.

In line with the spirit of CIBB, the 2016 meeting brought together researchers from different communities who address problems from different, but connected and often overlapping, perspectives. CIBB 2016 tackled the difficult task of bridging different backgrounds by providing an inclusive venue to discuss advances and future perspectives in different areas. It also fostered interaction between theory and practice, addressing both the theories underpinning the methodologies used to model and analyze biological systems, the practical applications of such theories, and the supporting technologies. Accordingly, participants at CIBB 2016 came from mathematical, computational, and medical backgrounds and institutions, both from academia and the private sector, offering collaboration opportunities and novel results in the areas of computational life sciences.

CIBB 2016 also offered a view on emerging and strongly developing trends and future opportunities at the edge of mathematics, computer and life sciences, such as synthetic biology, statistical investigation of genomic data, and applications to the understanding of complex diseases, such as cancer, and therapy opportunities. Along these lines, six keynote speakers, prominent scholars in their fields, presented the latest advances of their research within the context of their area of interest, and provided insights into open problems and future directions of general interest for the field. While papers in the main conference track addressed a rich set of open problems at the forefront of current research, the conference hosted six further special sessions on specific themes: biomedical databases, synthetic cell biology, high-performance computing in genetics, modelling for systems biology and medicine, survival analysis, and statistical inference in biological models. Researchers from Europe, Asia, USA, and Africa attended the conference. CIBB 2016 was made possible by the efforts of the Organizing, Program, and Steering Committees and by the support of sponsors and participants. CIBB 2016 was held in Stirling, UK, during September 1–3, 2016 (http://www.cs.stir.ac.uk/events/cibb2016/). With the continued support of the community, the next edition of CIBB will be held in Cagliari, Italy.

Overall, 61 contributions were submitted for consideration to CIBB 2016, amongst which 49 were invited for an oral presentation at the conference, after a first round of reviews (at this stage, each paper received an average of 3.7 reviews from the Program Committee and about 30 additional referees). Following the conference, selected papers were invited for further submission, after feedback and discussion from the conference.

This volume collects the papers that were accepted after a further round of reviews (2.5 for each paper, on average).

From 2004 to 2007, CIBB had the format of a special session of larger conferences, namely, WIRN 2004 in Perugia, WILF 2005 in Crema, FLINS 2006 in Genoa, and WILF 2007 in Camogli. Given the great success of the special session at WILF 2007 that included 26 strongly rated papers, the Steering Committee decided to turn CIBB into an autonomous conference starting with the 2008 edition in Vietri. The following editions in Italian venues were held in Genoa (2009), Palermo (2010), and Gargnano (2011). Until 2012, CIBB meetings were held annually in Italy with an increasing number of participants. CIBB 2012 was the first edition organized outside Italy, in Houston, then in Nice, France (2013), Cambridge, UK (2014), and Naples, Italy (2015). A rigorous peer-review selection process is applied every time to ultimately select the papers included in the program of the conference, in the conference proceedings published in the LNBI-LNCS book series by Springer, and in some cases, selected papers were published in special issues of well-qualified international journals, such as *BMC Bioinformatics*.

June 2017

Andrea Bracciali
Giulio Caravagna
David Gilbert
Roberto Tagliaferri

Organization

CIBB 2016 was jointly organized by the Computing Science and Mathematics, University of Stirling, UK, and the Dipartimento di Informatica, University of Salerno, Italy.

General Chairs

Andrea Bracciali	University of Stirling, UK
David Gilbert	Brunel University London, UK
Gilbert MacKenzie	UL, ENSAI and Keele, UK

Biostatistics Technical Chair

Marco Bonetti	Bocconi University, Italy

Bioinformatics Technical Chair

Ivan Merelli	CNR-ITB, Italy

Keynote Speakers

Mark Beaumont	School of Biological Sciences, Bristol University, UK
Natalio Krasnogor	School of Computing Science, Newcastle University, UK
Antonietta Mira	IDIDS, Università della Svizzera italiana and Università dell'Insubria, Italy
Bud Mishra	Courant Institute of Mathematical Sciences, NYU School of Medicine, NYU Mt. Sinai School of Medicine, USA
Daniela Paolotti	Institute for Scientific Interchange, Turin, Italy
Guido Sanguinetti	School of Informatics, University of Edinburgh, UK

Organizing Committee

Andrea Bracciali	University of Stirling, UK
Giulio Caravagna	Edinburgh University, UK
Muhamed Wael Farouq	University of Stirling, UK
Leslie Smith	University of Stirling, UK
Grace McArthur	University of Stirling, UK
Gemma Gardiner	University of Stirling, UK

Organizing Institutions

Publicity Chair

Francesco Masulli   University of Genova, Italy

Special Session and Tutorial Chair

Claudia Angelini   IAC-CNR, Italy

Publication Chair

Riccardo Rizzo   ICAR-CNR, Italy

Steering Committee

Pierre Baldi   University of California, Irvine, CA, USA
Elia Biganzoli   University of Milan, Italy
Clelia Di Serio   University Vita-Salute San Raffaele, Italy
Alexandru Floares   Oncological Institute Cluj-Napoca, Romania
Jon Garibaldi   University of Nottingham, UK
Nikola Kasabov   Auckland University of Technology, New Zealand
Francesco Masulli   University of Genova, Italy and Temple University,
                              PA, USA
Leif Peterson   TMHRI, Houston, Texas, USA
Roberto Tagliaferri   University of Salerno, Italy

Program Committee

Antonino Abbruzzo   University of Palermo, Italy
Federico Ambrogi   University of Milan, Italy
Claudio Angione   Teesside University, UK
Sansanee Auephanwiriyakul Chiang Mai University, Thailand
Krzysztof Bartoszek   Uppsala Universitet, Sweden
Gilles Bernot   University of Nice-Sophia Antipolis, France
Matteo Brilli   University of Padova, Italy
Giulio Caravagna   University of Edinburgh, UK
Davide Chicco   University of Toronto, Canada
Luisa Cutillo   University of Sheffield, UK
Angelo Facchiano   CNR, Italy
Enrico Formenti   University of Nice-Sophia Antipolis, France
Leonardo Franco   Universidad de Malaga, Spain
Christoph Friedrich   University of Dortmund, Germany
Marco Grzegorczyk   University of Groningen, The Netherlands
Maria Luisa Guerriero   Astrazeneca, UK

Saman K. Halgamuge	University of Melbourne, Australia
Sean Holden	University of Cambridge, UK
Yin Hu	Sage Bionetworks, USA
Antonella Iuliano	IAC-CNR, Italy
Pawel Labaj	BOKU Vienna, Austria
Paulo Lisboa	Liverpool John Moores University, UK
Giosué Lo Bosco	University of Palermo, Italy
Hassan Mahmoud	University of Genoa, Italy
Anna Marabotti	University of Salerno, Italy
Elena Marchiori	Radboud University, The Netherlands
Giancarlo Mauri	University of Milan, Italy
Luciano Milanesi	ITB-CNR, Italy
Dimitrios Milios	Edinburgh University, UK
Marianna Pensky	University of Central Florida, USA
Luca Pinello	Harvard University, USA
Vilda Purutçuoğlu	Middle East Technical University, Turkey
Daniele Ramazzotti	Stanford University, USA
Davide Risso	University of California, Berkeley, USA
Riccardo Rizzo	ICAR-CNR, Italy
Paolo Romano	IRCCS AOU San Martino IST, Genoa, Italy
Simona Rombo	University of Palermo, Italy
Stefano Rovetta	University of Genoa, Italy
Francesco Stingo	UT MD Anderson Cancer, USA
Paolo Tieri	IAC-CNR, Italy
Alfonso Urso	ICAR-CNR, Italy
Filippo Utro	IBM, USA
Alfredo Vellido	Universidad Politècnica de Catalunya, Spain
Veronica Vinciotti	Brunel University, UK
Blaz Zupan	University of Ljubljana, Slovenia

Special Session Organizers

Maria Raposo	Universidade Nova de Lisboa, Portugal
Quirina Ferreira	Universidade de Lisboa, Portugal
Andrea Antunes	Universidade Federal de Uberlândia, Brazil
Patricia Targon	Campana Universidade de São Paulo, Brazil
Rosalba Giugno	University of Verona, Italy
Giosue' Lo Bosco	University of Palermo, Italy
Alfredo Pulvirenti	University of Catania, Italy
Riccardo Rizzo	CNR-ICAR, Palermo, Italy
Paolo Cazzaniga	University of Milano-Bicocca, Italy
Marco S. Nobile	University of Milano-Bicocca, Italy
Chiara Damiani	University of Milano-Bicocca, Italy
Riccardo Colombo	University of Milano-Bicocca, Italy
Giancarlo Mauri	University of Milano-Bicocca, Italy
Zakaria Benmounah	Constantine 2 University, Algeria

Filippo Spiga	University of Cambridge, UK
Fabio Tordini	University of Turin, Italy
Dirk Husmeier	University of Glasgow, UK
Maurizio Filippone	Eurecom, France
Simon Rogers	University of Glasgow, UK
Mu Niu	University of Glasgow, UK
Benn Macdonald	University of Glasgow, UK
Elia Biganzoli	University of Milan, Italy
Clelia Di Serio	Vita-Salute San Raffaele University, Italy

Additional Reviewers

Rino Bellocco	Antonella Galizia	Guilherme Peretti Pezzi
Francesco Billari	Elisabeth Larsson	Leonardo Rundo
Marco Carone	Michele Loreti	Heba Saadeh
Enrico Cataldo	Giuseppe Marano	Andrea Tangherloni
Pietro Coretto	Paolo Milazzo	Claudia Tarantola
Federica Chiappori	Antonietta Mira	Paola Vicard
Eugenio Del Prete	Roberto Marangoni	Paolo Viviani
Maurizio Drocco	Rohan Nanda	Elisabeth Yaneske
Giulio Ferrero	Annalisa Orenti	Federico Zambelli

CIBB 2016 Sponsors

http://www.sicsa.ac.uk/

Predictive mOdelling for hEalthcare technology through MathS
http://poems.group.shef.ac.uk/

Italian Region of the IBS
http://ibs-italy.org/wps/wcm/
?page_id=363&lang=en

http://www.dondena.unibocconi.it/connect/
Cdr/Centro_Dondena/Home

http://www.cs.stir.ac.uk/

http://www.inns.org/bio_int

http://my.stirling.gov.uk/

This conference was organized with the support of EasyChair conference system.

Keynote Abstracts

Accelerating Synthetic Biology via Software and Hardware Advances

Natalio Krasnogor

School of Computing Science, Newcastle University

In this talk I will discuss recent work done in my lab that contributes towards accelerating the

$$\text{specify} \rightarrow \text{design} \rightarrow \text{model} \rightarrow \text{build} \rightarrow \text{test \& iterate}$$

biological engineering cycle. This will describe advances in biological programming languages for specifying combinatorial DNA libraries, the utilisation of off-the-shelf microfluidic devices to build the DNA libraries as well as data analysis techniques to accelerate computational simulations.

Professor Natalio Krasnogor is Professor of Computing Science and Synthetic Biology, co-directs Interdisciplinary Computing and Complex Biosystems (ICOS) research group and is director of Newcastle's Centre for Synthetic Biology and the Bioeconomy (CSBB). Krasnogor holds an EPSRC Leadership Fellowship in Synthetic Biology (SB) (EP/J004111/1 - 1.1M), is the PI on the EPSRC programme grant "Synthetic Portabolomics" (EP/N031962/1 - 4.3M) and is the overall lead in the EPSRC Synthetic Biology ROADBLOCK (EP/I031642/1, EP/I031812/1, EP/I03157X/1 - 1.7M) project involving Newcastle, Nottingham, Sheffield, Warwick and Bradford Universities. With expertise in Synthetic Biology, complex systems and machine intelligence, he attracted 4.5M as PI from UKRC and EU. Krasnogor gave several keynote talks (e.g. IEEE CEC, PPSN, GECCO, etc); has 160+ publications (H-index 36), with many papers in the top 0.1% and 1% for number of citations in computing science and also papers in Nature Biotech, Nature Chemistry, PNAs, NAR, EMBO Journal, etc. He won several best papers prizes as well as Bronze, Silver and Gold awards of the American Computing Society's (ACM) HUMIES and ACM's Impact award. From 2012 to 2014 he was the Science Director of the European Centre for Living Technologies (Italy) and was distinguished visiting professor at Ben Gurion University (Israel) in 2009 and Weizmann Institute of Science (Israel) in 2010, 2012 and 2013. Krasnogor current interests in Synthetic Biology are in the development of artificial intelligence techniques, including data analytics, the design of programming languages for biological engineering and research at the interface of nano and bio technology.

(Cancer) Genomics via (Sub)Optical Mapping

Bud Mishra

Computer Science, Mathematics, Engineering and Biology, Courant Institute,
Tandon School of Engineering, and NYU School of Medicine

The dream of a powerful integrated computational framework, only hinted at in Ibn Sina's Canon, can now be fulfilled at a global scale as a result of many recent advances: foundational advances in statistical inference; hypothesis-driven experiment design and analysis and the dissemination of peer-reviewed publications among communities of scientists; distributed large-scale databases of scientific and auxiliary experimental data; algorithmic approaches to model building and model checking; machine learning approaches to generate large number of hypotheses, and multiple hypotheses testing to tame computational complexity and false-discovery rates, etc. We will focus on an application centered on cancer - "the emperor of all maladies."

The topics this talk will cover include:

- Probabilistic causation
- Causal analysis of Cancer genome data
- Kernel based methods for survival analysis
- Improved single-cell/single-molecule data via SubOptical Mapping
- CHA and Therapy design
- Immuno-therapy
- Liquid Biopsies

Professor Bud Mishra is an American-Indian technologist, educator and mentor. He is currently a professor of computer science and mathematics at NYU's Courant Institute of Mathematical Sciences, professor of engineering at NYU's Tandon School of engineering, professor of human genetics at Mt. Sinai School of Medicine, visiting scholar at Cold Spring Harbor Laboratory and a professor of cell biology at NYU School of Medicine. Prof. Mishra has a degree in Science from Utkal University, in Electronics and Communication Engineering from IIT, Kharagpur, and MS and PhD degrees in Computer Science from Carnegie-Mellon University. He has advisory experience in Computer and Data Science (ATTAP, brainiad, Genesis Media, Pypestream, and Tartan Laboratories), Finance (Instadat, PRF, LLC, and Tudor Investment), Robotics and Bio- and Nanotechnologies (Abraxis, Bioarrays, InSilico, MRTech, OpGen and Seqster). He has advised and mentored more than 35 graduate students and post-docs in the areas of computer science, robotics and control engineering, applied mathematics, finance, biology and medicine. He holds 21 issued and 23 pending patents in areas ranging over robotics, model checking, intrusion detection, cyber security, emergency response, disaster management, data analysis, biotechnology, nanotechnology, genome mapping and sequencing, mutation calling, cancer biology, fintech, adtech, internet architecture and linguistics. His pioneering work includes: first application of model checking to hardware verification; first robotics technologies for grasping, reactive grippers and work holding; first single molecule genotype/haplotype mapping technology (Optical Mapping); first analysis of copy number variants with a segmentation algorithm, first whole-genome haplotype assembly technology (SUTTA), first clinical-genomic variant/base calling technology (TotalRecaller), and current work in progress continuing in the areas of liquid biopsies, cancer data, cyber security, cryptography, financial engineering and internet of the future. He is a fellow of IEEE, ACM and AAAS, a Distinguished Alumnus of IIT-Kgp, and a NYSTAR Distinguished Professor.

Statistical Inference on Large-Scale Gene Duplication Networks

Antonietta Mira

IDIDS, Università della Svizzera italiana and Università dell'Insubria

Many systems of scientific interest can be investigated as networks, where network nodes correspond to the elements of the system and network edges to interactions between the elements. Increasing availability of large-scale biological data and steady improvements in computational capacity are continuing to fuel the growth of this field. Network models are now used commonly to investigate biological complexity at the systemic level. Gene duplication is one of the main drivers of the evolution of genomes, and network models based on gene duplication were one of the first large-scale models used in systems biology. An attractive feature of some of these so-called duplication-divergence models is their analytical tractability, but there is typically no statistically principled way to estimate their model parameters from empirical data. This is a reflection of a more general divide between the two prominent paradigms to the modeling of networks, which are the approach of mechanistic networks models and the approach of statistical network models. Mechanistic network models assume that the microscopic mechanisms governing network formation and evolution at the level of individual nodes are known, and questions often focus on understanding macroscopic features that emerge from repeated application of these known mechanisms. The statistical approach, in contrast, often starts from observed network structures and attempts to infer some aspects about the underlying data generating process. Mechanistic network models provide insight into how the network is formed and how it evolves at the level of individual nodes, but as mechanistic rules typically lead to complex network structures, it is difficult to assign a probability to any given network realizations that a mechanistic model may generate. Because of this difficulty, there is typically no closed form expression for likelihood for these models and, consequently, likelihood based inference for learning from data is not possible. We have developed a principled statistical framework, based on Approximate Bayesian Computation, to bring some of the mechanistic network models into the realm of statistical inference. This approach is feasible because given a set of parameter values, it is easy to sample network configurations from most mechanistic models. I will introduce this general framework and demonstrate its application to large-scale gene duplication networks, where it can be used to infer model parameters, and their associated uncertainties, for mechanistic network models from empirical data.

Joint work with Jukka-Pekka Onnela, Department of Biostatistics, Harvard

Antonietta Mira is a professor of statistics, and co-founder and co-director of the InterDisciplinary Institute of Data Science, IDIDS, at Universit della Svizzera italiana, where she also served as the Vice-Dean in the Faculty of Economics (2013–2015). She is a fellow of the International Society for Bayesian Analysis, a member of the Istituto Lombardo Accademia di Scienze e Lettere, a visiting fellow of the Isaac Newton Institute for Mathematical Sciences at Cambridge University (2014 and 2016), and of the Queensland University of Technology (2016–2019). She is the principal investigator on several projects at the Swiss National Science Foundation and a member of multiple scientific committees representing her areas of expertise: Bayesian statistical models and efficient Monte Carlo simulation algorithms and theory. She has been member of the board of the ISBA Section on Bayesian Computation since 2013 and has been member of the scientific program committee and of the organizing committee of the joint international meeting of the Institute of Mathematical Statistics/International Society for Bayesian Analysis, aka MCMSki. Her current research focuses on data science and methodological and computational statistics, both of which have a clear interdisciplinary scope across social science, biology, genetics, economics and finance. She is often invited to talk at international scientific conferences. She served on the editorial board of Bayesian Analysis, Statistica Sinica and the Journal of Computational and Graphical Statistics as has been Chief Guest Editor for two special issues of Statistics and Computing. She has been involved in public engagement (such as EXPO Milano 2015), has delivered public lectures on several science festivals, and is the scientific lead for the exhibit Numbed by Numbers! Antonietta holds a PhD in Computational Statistics (1998, University of Minnesota, US) and a Doctorate in Methodological Statistics (1995, University of Trento, Italy). She has earned her Bachelor1s in Economics, summa cum laude, from the University of Pavia, Italy. Her work has been published in over 60 scientific articles and she is co-author of the book Mathematical-Magic (Aboca, 2012).

Contents

Module Detection Based on Significant Shortest Paths
for the Characterization of Gene Expression Data 1
   Daniele Pepe

Information-Theoretic Active Contour Model for Microscopy
Image Segmentation Using Texture 12
   Veronica Biga and Daniel Coca

Host Phenotype Prediction from Differentially Abundant Microbes
Using RoDEO 27
   Anna Paola Carrieri, Niina Haiminen, and Laxmi Parida

DeepScope: Nonintrusive Whole Slide Saliency Annotation
and Prediction from Pathologists at the Microscope 42
   Andrew J. Schaumberg, S. Joseph Sirintrapun, Hikmat A. Al-Ahmadie,
   Peter J. Schüffler, and Thomas J. Fuchs

PLS-SEM Mediation Analysis of Gene-Expression Data for the Evaluation
of a Drug Effect 59
   Daniele Pepe and Tomasz Burzykowski

A Novel Algorithm for CpG Island Detection in Human Genome Based
on Clustering and Chaotic Particle Swarm Optimization 70
   Abdelbasset Boukelia, Zakaria Benmounah, Mohamed Batouche,
   Bouchera Maati, and Ikram Nekkache

COSYS: A Computational Infrastructure for Systems Biology 82
   Fabio Cumbo, Marco S. Nobile, Chiara Damiani, Riccardo Colombo,
   Giancarlo Mauri, and Paolo Cazzaniga

Statistical Texture-Based Mapping of Cell Differentiation
Under Microfluidic Flow 93
   Veronica Biga, Olívia M. Alves Coelho, Paul J. Gokhale,
   James E. Mason, Eduardo M.A.M. Mendes, Peter W. Andrews,
   and Daniel Coca

Constraining Mechanism Based Simulations to Identify Ensembles
of Parametrizations to Characterize Metabolic Features 107
   Riccardo Colombo, Chiara Damiani, Giancarlo Mauri,
   and Dario Pescini

Process Algebra with Layers: Multi-scale Integration Modelling Applied
to Cancer Therapy. 118
   Erin Scott, James Nicol, Jonathan Coulter, Andrew Hoyle,
   and Carron Shankland

A Problem-Driven Approach for Building a Bioinformatics GraphDB 134
   Antonino Fiannaca, Massimo La Rosa, Laura La Paglia,
   Antonio Messina, Riccardo Rizzo, and Alfonso Urso

Parameter Inference in Differential Equation Models of Biopathways
Using Time Warped Gradient Matching. 145
   Mu Niu, Simon Rogers, Maurizio Filippone, and Dirk Husmeier

IRIS-TCGA: An Information Retrieval and Integration System
for Genomic Data of Cancer. 160
   Fabio Cumbo, Emanuel Weitschek, Paola Bertolazzi,
   and Giovanni Felici

Effect of UV Radiation on DPPG and DMPC Liposomes in Presence
of Catechin Molecules. 172
   Filipa Pires, Gonçalo Magalhães-Mota, Paulo António Ribeiro,
   and Maria Raposo

Inference in a Partial Differential Equations Model of Pulmonary
Arterial and Venous Blood Circulation Using Statistical Emulation 184
   Umberto Noè, Weiwei Chen, Maurizio Filippone, Nicholas Hill,
   and Dirk Husmeier

Ensemble Approaches for Stable Assessment of Clusters
in Microbiome Samples. 199
   Sanja Brdar and Vladimir Crnojević

Multilayer Data and Document Stratification for Comorbidity Analysis 209
   Kevin Heffernan, Pietro Liò, and Simone Teufel

Evolving Dendritic Morphologies Highlight the Impact
of Structured Synaptic Inputs on Neuronal Performance. 220
   Mohammad Ziyad Kagdi

Semantic Clustering for Identifying Overlapping Biological Communities . . . 235
   Hassan Mahmoud, Francesco Masulli, and Stefano Rovetta

Author Index 249

Module Detection Based on Significant Shortest Paths for the Characterization of Gene Expression Data

Daniele Pepe[✉]

I-BioStat, Hasselt University, Campus Diepenbeek, Hasselt, Belgium
daniele.pepe@uhasselt.be

Abstract. The characterization of diseases in terms of perturbed gene modules was recently introduced for the analysis of gene expression data. Some approaches were proposed in literature, but most of them are inductive approaches. This means that they try to infer key gene networks directly from data, ignoring the biological information available. Here a unique method for the detection of perturbed gene modules, based on the combination of data and hypothesis-driven approaches, is described. It relies upon biological metabolic pathways and significant shortest paths evaluated by structural equation modeling (SEM). The procedure was tested on a microarray experiment concerning tuberculosis (TB) disease. The validation of the final disease module was principally done by the Wang similarity semantic index and the Disease Ontology enrichment analysis. Finally, a topological analysis of the module via centrality measures and the identification of the cut vertices allowed to unveil important nodes in the disease module network. The results obtained were promising, as shown by the detection of key genes for the characterization of the studied disease.

Keywords: Disease module · Structural equation modeling · Gene expression data · Significant shortest paths

1 Introduction

The reductionist approach in medicine, based on the principle of "divide and conquer", although useful, present limits when it is necessary to explain the onset and progression of complex diseases. In fact, the approach is rooted in the assumption that if a complex problem is divided into more understandable and smaller units, then by their reconstruction, it is possible to unveil the studied complex problem.

For this reason, there are lists of genes associated with diseases. OMIM, a free database [1], for example, offers a catalogue of genes with the relative description of their role in the associated phenotypes. Conversely to this point of view, in the 1972, Anderson, in the article "More is complex" [2] affirms that the behaviour of large and complex aggregate of elementary particles cannot be understood in terms of a simple extrapolation of the properties of a few particle. At each level of complexity entirely new properties appear. For this reason, we are assisting to the passage from the reductionist approach to the systemic approach [3]. Most of biological networks are

© Springer International Publishing AG 2017
A. Bracciali et al. (Eds.): CIBB 2016, LNBI 10477, pp. 1–11, 2017.
DOI: 10.1007/978-3-319-67834-4_1

subjected to specific laws [4], as the small world phenomena, which affirms that there are relatively short paths between any pair of nodes, the scale-free principle, with the consequence that there are few highly connected nodes; the local hypothesis i.e. the presence of modules, highly interlinked local regions in the network in which the components are involved in same biological processes.

The last property, modularity, is a general design principle in biological systems and has been observed also in transcriptional regulation networks [5]. In biology, modularity refers to a group of physically or functionally linked molecules (nodes) that work together to achieve a (relatively) distinct function.

Applying module level analysis should help to study biological systems at different levels and to understand which properties characterize the level of complexity considered. Many approaches exist that use a gene-module view as the basic building blocks of the analysis [6]. In general, it is used to divide the module identification in three main approaches: (1) network-based approach; (2) expression-based approach; (3) pathway-based approach [7]. The first approach is based on the topology of network, and modules are defined as subsets of vertices with high connectivity between them and less with external nodes. The second approach uses gene expression data for inferring modules of genes exhibiting similar expression by, for example, clustering methods. The third approach detects expression changes in biological pathways, group of genes that accomplishes specific biological functions.

The approach proposed in this paper, it is a mix and more general approach that takes advantage of the three approaches previously described. In fact, the pathway-based approach was used to detect perturbed KEGG pathways, then the network-approach was employed to identify the shortest paths between the differentially expressed genes (DEGs) and finally significant shortest paths (SSPs) were found using the expression data and structural equation modeling (SEM) [8].

The idea to consider shortest paths between DEGs to understand how they are connected is not new [9–11]. However, differently from the methods previously proposed, the key elements to test are constituted by shortest paths got from the network generated by the fusion of the relevant pathways. In this way, it is possible to consider the inter-pathway connectivity of DEGs by significant shortest paths tested by multiple group SEM. All the SSPs were joined to have the final perturbed disease module.

2 Materials and Methods

The classical differential gene expression analysis allows to identify DEGs. The differential analyses at gene level were performed by Significance Analysis of Microarray (SAM) [12], but any other procedure can be used. The next step is to find the network context where the DEGs act. The classical way is by pathway analysis also if it is not always able to detect the required information. In this situation, a solution could be take all the pathways containing at least one DEG. In the tuberculosis case the "Signaling Impact Pathway Analysis" (SPIA) was applied [13]. The corresponding perturbed KEGG pathways can be represented as mixed graphs, where the nodes represent genes and the edges represent multiple functional relationships between genes as activation, inhibition, binding etc. The core idea for building a disease module is to understand how the DEGs, in the perturbed pathways, are connected between them. The first step for reaching this goal is to merge all the relevant pathways in a unique graph. The second step is to find the significant shortest paths that put in communication every couple of DEGs.

Each shortest path could be represented as a list of nodes $P = (p_i, p_{i+1},..., p_{j-1}, p_j)$ and a list of the corresponding edges $E = (e_{i(i+1)}, ..., e_{(j-1)j})$ where (p_i, p_j) are DEGs and $(p_{i+1},..., p_{j-1})$ can be DEGs or other microarray genes. In this analysis, every edge is directed. A shortest path can be codified as a structural equation (SE) model, in the following way:

$$P_j = \beta_{ji}P_i + E_j \tag{1}$$

where P_j represents every gene in the path that is influenced directly by the gene P_i; β_{ji} is the strength of relationship between node P_i and P_j; E_j is a term that represent external causes that have an effect on P_j but not explicated in the model. Considering that the shortest paths selected are induced paths, every shortest path can be represented by $j - 1$ simple linear equations, where j is the number of nodes in the path.

For the estimation of the parameters β_{ij}, the Maximum Likelihood estimation (MLE) is used, assuming that all observed variables have a multinormal distribution. For finding the SSPs, the following omnibus test was performed:

$$H_0 : \sum{}_1 (\theta) = \sum{}_2 (\theta) \, vs. \, H_1 : \sum{}_1 (\theta) \neq \sum{}_2 (\theta) \tag{2}$$

where $\sum_1(\theta)$ and $\sum_2(\theta)$ are the model-implied covariance matrices of the groups one and two, and θ represents the parameter of the model. The test verifies if the difference between the model-implied covariance matrices of each group are statistically significant (H_1) or not (H_0) The statistical significance is determined by comparison of likelihood ratio test (LRT) chi-square ($\chi^2 diff$) values at a given degree of freedom (d.f. ($\chi^2 diff$)). If there is a significant difference ($P < 0.05$), after the Benjamin-Hochberg correction in the chi-squared goodness-of-fit index, the shortest path is considered statistically significant. All the SSPs were merged to obtain the final disease module. The final module is a weighted graph, where the weights correspond to the parameters estimated by SEM. The weights were used for the topological analysis, subsequently described. To validate the procedure two different approaches were employed: (1) enrichment analysis based on Disease Ontology (DO), to verify if the module genes are associated to the family of diseases connected with the analysed disease; (2) semantic similarity index, based on DO terms, between the list of genes associated "a priori" to the disease and the list of genes present in the module. DO creates a single structure for the classification of disease and permits to represent them in a relational ontology [14]. For the semantic similarity, the graph-based strategy proposed by Wang et al. [15] was applied. The similarity goes from 0, when the lists of genes are not associated, to 1, when the lists of genes contribute to the same DO terms. For finding the a priori genes, a search on Entrez Gene [16] was done. For each set of genes, an enrichment analysis on DO was done and the list of enriched terms was compared with those obtained by the genes in the module. Finally, a basic network analysis was performed based on measures of centralities (betweenness and the Bonacich power centrality score) and connectivity (detection of articulation nodes). The betweenness centrality for the gene v is defined as:

$$\sum_{i \neq j, i \neq v, j \neq v} g_{ivj} \backslash g_{ij} \tag{3}$$

where g_{ivj} are the shortest paths between the node i and j in which the node v is present, while g_{ij} is the number of all shortest paths between the nodes i and j. The normalized betweenness centrality is obtained:

$$B_h = 2B/\left(n^2 - 3n + 2\right) \tag{4}$$

where B is the raw betweenness and n the number of nodes. The Bonacich power measure corresponds to the notion that the power of a vertex is recursively defined by the sum of the power of its alters. The formula is the following:

$$C(\alpha, \beta) = \alpha \left(I - \beta R\right)^{-1} R\mathbf{1} \tag{5}$$

where α is a scaling vector, β is an attenuation factor to weight the centrality of the nodes, R is the adjacency matrix, I is the identity matrix, $\mathbf{1}$ is a matrix of all ones. The articulation nodes are the minimum set of nodes which removal increases the number of connected components. They represent the nodes that allow to have a connected module.

The procedure was tested on the dataset GSE54992, where the group of samples of active tuberculosis (TB) (9 samples) were compared to healthy control (6 samples).

3 Results

SAM revealed 2152 significant genes using as delta value a value of 1.178 corresponding to a FDR value less than 0.05 and a minimum fold change of 2. On the 2152 DEGs, SPIA revealed important perturbed pathways (see Table 1), most of them associated to inflammation and infection as the cytokine-cytokine receptor interaction, the chemokine signaling pathway, the NF-κB signaling pathway, Legionellosis, Malaria. The first two pathways for example are induced in the lung in the response to TB infection to accumulate and mediate formation of granulomas, bacterial control and protection against the infection [17]. The pathways were transformed in graph and subsequently merged.

The next step was to find the shortest paths between every couple of DEGs on the total graph. The total number of shortest paths resulted of 1493 with 316 genes and 745 connections involved. For each shortest path a structural model was generated and tested to detect those significant. 260 out of 1493 were found relevant involving 206 genes and 330 connections. Figure 1 shows the graph obtained by the fusion of the 260 significant paths.

Table 1. Significant perturbed pathways for tuberculosis data by SPIA analysis.

Name	pSize	NDE	pNDE	pPERT	pGFdr
Cytokine-cytokine receptor interaction	241	66	0,000	0,000	0,000
Chemokine signaling pathway	177	42	0,000	0,000	0,000
NF-kappa B signaling pathway	75	23	0,000	0,021	0,000
Osteoclast differentiation	120	27	0,000	0,001	0,000
Legionellosis	39	16	0,000	0,287	0,000
Complement and coagulation cascades	65	22	0,000	0,882	0,000
Staphylococcus aureus infection	26	11	0,000	0,035	0,000
Proteoglycans in cancer	198	36	0,000	0,007	0,001
Rheumatoid arthritis	17	7	0,001	0,014	0,003
Pathways in cancer	308	49	0,001	0,011	0,003
Inflammatory mediator regulation of TRP channels	87	18	0,003	0,009	0,004
Toxoplasmosis	91	22	0,000	0,510	0,007
Focal adhesion	206	32	0,011	0,006	0,008
MAPK signaling pathway	248	40	0,003	0,051	0,014
Viral myocarditis	26	9	0,001	0,219	0,016
Mineral absorption	8	5	0,000	0,344	0,016
Pertussis	49	14	0,000	0,630	0,016
Systemic lupus erythematosus	12	4	0,028	0,007	0,016
ECM-receptor interaction	86	14	0,054	0,004	0,017
Intestinal immune network for IgA production	25	9	0,001	0,441	0,017
Leishmaniasis	47	12	0,002	0,115	0,018
Influenza A	105	21	0,002	0,143	0,019
Amoebiasis	45	13	0,000	0,740	0,019
Rap1 signaling pathway	204	35	0,002	0,228	0,021
Toll-like receptor signaling pathway	97	21	0,001	0,900	0,032
Melanogenesis	99	15	0,079	0,008	0,032
Regulation of actin cytoskeleton	182	30	0,006	0,121	0,034
Malaria	11	5	0,003	0,266	0,037
Sphingolipid signaling pathway	98	20	0,002	0,483	0,041

pSize = number of genes in the pathway; **NDE** = number of DEGs in the pathway; **pNDE** = p-value for the enrichment analysis; **pPERT** = p-value for the accumulated perturbation; **pGFdr** = combined p-value from the two previuos two p-value adjusted for the Fdr.

For the validation of the module two different analyses were performed: (1) one looks for diseases enrichment analysis; (2) the other measures the semantic similarity between the genes associated "a priori" with TB and those in the module. The results of the enrichment analysis on DO were very interesting considering that in the list there of the enriched diseases there was TB and other diseases that share common biological mechanisms as cancer, Salmonella infection and pertussis (see Fig. 2).

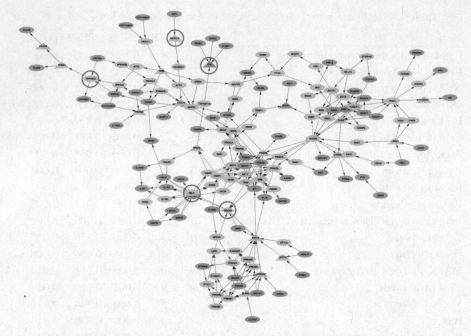

Fig. 1. TB disease module where the green nodes are the DEGs, the yellow ones the not DEGs that allows to the perturbation signal to propagate between the DEGs. Triangular nodes are the articulation point and the diamond node represents the gene SOCS3. Some key genes are highlighted with red circles. (Color figure online)

For the validation with the semantic similarity approach, the genes associated with TB in the database "Gene" of NCBI were detected. They were 312. The intersection of the genes in the module with those associated to TB revealed 24 genes in common. Of these, 13 were DEGs while the remaining genes were not DEGs. This result shows the limits of differential analysis when this stops to the detection of DEGs without understanding how these are connected. The semantic similarities between the two lists, based on DO terms was of 0.913. This means that the lists are highly related in terms of associated diseases. The following analyses were performed: (1) detection of the top 10 genes with the highest betweenness score and with the highest Bonacich's power centrality measure (Table 2); (2) detection of the cut vertices (Table 3). The highest values of normalized betweenness and Bonacich measure, on the weighted disease module, were associated to the gene SOCS3. This is very encouraging as it is known its fundamental role in immune responses to pathogens [18]. In fact, SOCS3 was considered as the most important family member for the association to autoimmunity, oncogenesis, diabetes and pathogenic immune evasion. It regulates both cytokine- and pathogen-induced cascades. Considering its importance, it was proposed as a therapeutic target [19]. Regarding the articulation point analysis, many interesting genes are present as the TNF, whose inhibition increases the risk of infections [20], as well as MAPK12 and some relevant kinases [21].

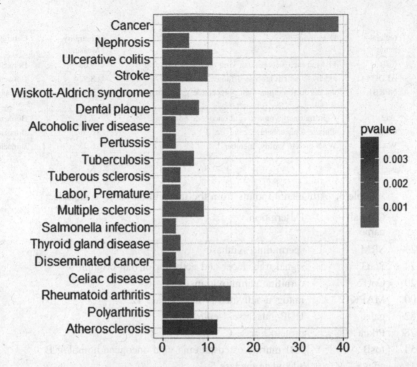

Fig. 2. Enriched diseases using the gene in the SSP module. Tuberculosis in one of the most enriched disease.

Table 2. Most relevant tuberculosis genes in the disease module based on betweeness (Betw) and Bonacich centrality measures. ·

Entrez	Official name	Description	Centrality Measure	Centrality
9021	socs3	suppressor of cytokine signaling 3	0,06	Betw
4790	NFKB1	nuclear factor kappa light polypeptide gene enhancer in B-cells 1	0,05	Betw
8517	ikbkg	inhibitor of kappa light polypeptide gene enhancer in B-cells, kinase gamma	0,05	Betw
998	Cdc42	cell division cycle 42 (GTP binding protein, 25 kDa);	0,04	Betw
8660	irs2	insulin receptor substrate 2	0,03	Betw
4301	mllt4	myeloid/lymphoid or mixed-lineage leukemia	0,02	Betw
5294	PIK3CG	phosphoinositide-3-kinase, catalytic, gamma polypeptide	0,02	Betw
8826	IQGAP1	IQ motif containing GTPase activating protein 1	0,02	Betw
25945	PVRL3	poliovirus receptor-related 3	0,02	Betw
6714	Src	v-src sarcoma (Schmidt-Ruppin A-2) viral oncogene homolog	0,02	Betw
9021	socs3	suppressor of cytokine signaling 3	20,25	Bonacich
4301	mllt4	myeloid/lymphoid or mixed-lineage leukemia	14,24	Bonacich
5209	pfkfb3	6-phosphofructo-2-kinase/fructose-2,6-biphosphatase 3	12,24	Bonacich
10458	BAIAP2	BAI1-associated protein 2	10,91	Bonacich

(continued)

Table 2. (*continued*)

Entrez	Official name	Description	Centrality Measure	Centrality
2354	fosB	FBJ murine osteosarcoma viral oncogene homolog B	9,88	Bonacich
8659	ALDH4A1	aldehyde dehydrogenase 4 family, member A1	8,92	Bonacich
4790	NFKB1	nuclear factor kappa light polypeptide gene enhancer in B-cells 1	7,74	Bonacich
208	akt2	v-akt murine thymoma viral oncogene homolog 2	7,43	Bonacich
92579	G6pc3	glucose 6 phosphatase, catalytic, 3	6,31	Bonacich
8936	WASF1	WAS protein family, member 1	6,25	Bonacich

Table 3. Articulation points from the tuberculosis disease module.

Entrez	Official name	Description
6723	SRM	spermidine synthase
6774	Stat3	signal transducer and activator of transcription 3
126129	Cpt1c	carnitine palmitoyltransferase 1C
6300	MAPK12	mitogen-activated protein kinase 12
2582	gale	UDP-galactose-4-epimerase
5578	Prkca	protein kinase C, alpha
2354	fosB	FBJ murine osteosarcoma viral oncogene homolog B
111	adcy5	adenylate cyclase 5
5332	Plcb4	phospholipase C, beta 4
27165	GLS2	glutaminase 2 (liver, mitochondrial)
7124	TNF	tumor necrosis factor (TNF superfamily, member 2)
217	ALDH2	aldehyde dehydrogenase 2 family (mitochondrial)
51005	AMDHD2	amidohydrolase domain containing 2
1573	cyp2j2	cytochrome P450, family 2, subfamily J, polypeptide 2
2744	GLS	glutaminase
3984	limk1	LIM domain kinase 1
7186	traF2	TNF receptor-associated factor 2
7358	UGDH	UDP-glucose dehydrogenase
1500	CTNND1	catenin (cadherin-associated protein), delta 1
2673	GFPT1	glutamine-fructose-6-phosphate transaminase 1
5563	prkaa2	protein kinase, AMP-activated, alpha 2 catalytic subunit
1994	ELAVL1	ELAV (embryonic lethal, abnormal vision, Drosophila)-like 1
4942	OAT	ornithine aminotransferase (gyrate atrophy)
8503	PIK3R3	phosphoinositide-3-kinase, regulatory subunit 3 (gamma)
5881	rac3	ras-related C3 botulinum toxin substrate 3
5743	PTGS2	prostaglandin-endoperoxide synthase 2
2043	EPHA4	EPH receptor A4
5742	Ptgs1	prostaglandin-endoperoxide synthase 1
5361	PLXNA1	plexin A1

(*continued*)

Table 3. (*continued*)

Entrez	Official name	Description
5567	PRKACB	protein kinase, cAMP-dependent, catalytic, beta
8660	irs2	insulin receptor substrate 2
55577	nagK	N-acetylglucosamine kinase
2773	GNAI3	guanine nucleotide binding protein (G protein)
26	ABP1	amiloride binding protein 1
5365	PLXNB3	plexin B3
1793	DOCK1	dedicator of cytokinesis 1
7132	TNFRSF1A	tumor necrosis factor receptor superfamily, member 1A
2805	GOT1	glutamic-oxaloacetic transaminase 1
8659	ALDH4A1	aldehyde dehydrogenase 4 family, member A1
4893	NRAS	neuroblastoma RAS viral (v-ras) oncogene homolog
5502	ppp1r1a	protein phosphatase 1, regulatory (inhibitor) subunit 1A
3065	Hdac1	histone deacetylase 1
2534	FYN	FYN oncogene related to SRC, FGR, YES
11069	Rapgef4	Rap guanine nucleotide exchange factor (GEF) 4
4790	NFKB1	nuclear factor of kappa light polypeptide gene enhancer in B-cells 1
5879	rac1	ras-related C3 botulinum toxin substrate 1
998	Cdc42	cell division cycle 42

4 Conclusion

In this paper, it was proposed a module level analysis for gene expression data that could take over the present methods for the identification of modules. It is used to divide the detection approaches in network-based, expression-based and pathway based. The approach here described is a mixed approach that starting from relevant pathways in which the DEGs are involved, detects the significant shortest paths by network, gene expression information and statistical analysis. The new concept is surely connected to the use of SEM for testing the significance of each shortest path model and the possibility to consider more pathways together, allowing to overcome the limiting idea of the pathway independence. Briefly, the pipeline consists in the following points: (1) discovering of DEGs associated to the disease; (2) understanding on which pathways the DEGs act; (3) joining in a unique graph all the relevant pathways; (4) performing the significant shortest path analysis for finding the disease module. The procedure was tested on a gene expression microarray concerning TB, but it can be applied to any gene expression experiment where the two-groups comparison is requested. The differential analysis of the shortest paths revealed significant shortest paths that characterize the experimental group on the control. The module obtained merging all the SSPs allowed to detect the key molecular network that could explain the disease. Very important genes were found as the SOCS3, TNF and MAPK2. The validation of the module by DO enrichment and similarity analysis has highlighted that the genes in the modules are strictly associated to the a priori genes connected with the disease. In conclusion, the approach, is surely notable as new approach for downstream analysis of gene expression data. Future developments could be the application of the procedure to data from the integration of different NGS experiments.

Funding acknowledgement. This research was funded by the MIMOmics grant of the European Union's Seventh Framework Programme (FP7-Health-F5-2012) under the grant agreement number 305280.

References

1. Hamosh, A., Scott, A.F., Amberger, J.S., Bocchini, C.A., McKusick, V.A.: Online Mendelian Inheritance in Man (OMIM), a knowledgebase of human genes and genetic disorders. Nucleic Acids Res. **33**(suppl 1), D514–D517 (2005). doi:10.1093/nar/gki033
2. Anderson, P.W.: More is different. Science **177**(4047), 393–396 (1972). doi:10.1126/science.177.4047.393
3. Ahn, A.C., Tewari, M., Poon, C.S., Phillips, R.S.: The limits of reductionism in medicine: could systems biology offer an alternative? PLoS Med **3**(6), e208 (2006). doi:10.1371/journal.pmed0030208
4. Barabási, A.L., Gulbahce, N., Loscalzo, J.: Network medicine: a network-based approach to human disease. Nat. Rev. Genet. **12**(1), 56–68 (2011). doi:10.1038/nrg2918
5. Girvan, M., Newman, M.: Community structure in social and biological networks. Proc. Natl. Acad. Sci. **99**(12), 7821–7826 (2002). doi:10.1073/pnas.122653799
6. Segal, E., Friedman, N., Kaminski, N., Regev, A., Koller, D.: From signatures to models: understanding cancer using microarrays. Nat. Genet. **37**, S38–S45 (2005). doi:10.1038/ng1561
7. Wang, X., Dalkic, E., Wu, M., Chan, C.: Gene module level analysis: identification to networks and dynamics. Curr. Opin. Biotechnol. **19**(5), 482–491 (2008). doi:10.1016/j.copbio.2008.07.011
8. Kline, R.B.: Principles and Practice of Structural Equation Modeling. Guilford Press (2011). doi:10.1111/insr.12011_25
9. Pepe, D., Grassi, M.: Investigating perturbed pathway modules from gene expression data via structural equation models. BMC Bioinform. **15**(1), 1–15 (2014). doi:10.1186/1471-2105-15-132
10. Pepe, D., Hwan, D.J.: Estimation of dysregulated pathway regions in MPP+ treated human neuroblastoma SH-EP cells with structural equation model. BioChip J. **9**(2), 131–138 (2015). doi:10.1007/s13206-015-9206-3
11. Pepe, D., Hwan, D.J.: Comparison of perturbed pathways in two different cell models for Parkinson's Disease with structural equation model. J. Comput. Biol. **23**(2), 90–101 (2016). doi:10.1089/cmb.2015.0156
12. Tusher, V.G., Tibshirani, R., Chu, G.: Significance analysis of microarrays applied to the ionizing radiation response. Proc. Natl. Acad. Sci. **98**(9), 5116–5121 (2001). doi:10.1073/pnas.091062498
13. Tarca, A.L., Draghici, S., Khatri, P., Hassan, S.S., Mittal, P., Kim, J.S., Kim, C.J., Kusanovic, J.P., Romero, R.: A novel signaling pathway impact analysis. Bioinformatics **25**(1), 75–82 (2009). doi:10.1093/bioinformatics/btn577

14. Schriml, L.M., Arze, C., Nadendla, S., Chang, Y.W.W., Mazaitis, M., Felix, V., Feng, G., Kibbe, W.A.: Disease ontology: a backbone for disease semantic integration. Nucleic Acids Res. **40**(D1), D940–D946 (2012). doi:10.1093/nar/gkr972
15. Wang, J.Z., Du, Z., Payattakool, R., Philip, S.Y., Chen, C.F.: A new method to measure the semantic similarity of GO terms. Bioinformatics **23**(10), 1274–1281 (2007). doi:10.1093/bioinformatics/btm087
16. Maglott, D., Ostell, J., Pruitt, K.D., Tatusova, T.: Entrez Gene: gene-centered information at NCBI. Nucleic Acids Res. **39**(suppl 1), D52–D57 (2011). doi:10.1093/nar/gkq1237
17. Slight, S.R., Khader, S.A.: Chemokines shape the immune responses to tuberculosis. Cytokine Growth Factor Rev. **24**(2), 105–113 (2013). doi:10.1016/j.cytogfr.2012.10.002
18. Carow, B., Reuschl, A.K., Gavier-Widén, D., Jenkins, B.J., Ernst, M., Yoshimura, A., Chambers, B.J., Rottenberg, M.E.: Critical and independent role for SOCS3 in either myeloid or T cells in resistance to Mycobacterium tuberculosis. PLoS Pathog. **9**(7), e1003442 (2013). doi:10.1371/journal.ppat.1003442
19. Mahony, R.A., Diskin, C., Stevenson, N.J.: SOCS3 revisited: a broad regulator of disease, now ready for therapeutic use? Cell. Molecular Life Sci. **1**(1), 1–14 (2016). doi:10.1007/s00018-016-2234-x
20. Sichletidis, L., Settas, L., Spyratos, D., Chloros, D., Patakas, D.: Tuberculosis in patients receiving anti-TNF agents despite chemoprophylaxis. Int. J. Tuberc. Lung Dis. **10**(10), 1127–1132 (2006)
21. Song, C.H., Lee, J.S., Lee, S.H., Lim, K., Kim, H.J., Park, J.K., Paik, T.H., Jo, E.K.: Role of mitogen-activated protein kinase pathways in the production of tumor necrosis factor-α, interleukin-10, and monocyte chemotactic protein-1 by Mycobacterium tuberculosis H37Rv-infected human monocytes. J. Clin. Immunol. **23**(3), 194–201 (2003)

Information-Theoretic Active Contour Model for Microscopy Image Segmentation Using Texture

Veronica Biga[1]([✉]) and Daniel Coca[2]

[1] Faculty of Biology, Medicine and Health, The University of Manchester,
Oxford Rd, M13 9PL Manchester, UK
veronica.biga@manchester.ac.uk
[2] Department of Automatic Control and Systems Engineering,
The University of Sheffield, Mappin Street, S1 3JD Sheffield, UK
d.coca@sheffield.ac.uk

Abstract. High throughput technologies have increased the need for automated image analysis in a wide variety of microscopy techniques. Geometric active contour models provide a solution to automated image segmentation by incorporating statistical information in the detection of object boundaries. A statistical active contour may be defined by taking into account the optimisation of an information-theoretic measure between object and background. We focus on a product-type measure of divergence known as Cauchy-Schwartz distance which has numerical advantages over ratio-type measures. By using accurate shape derivation techniques, we define a new geometric active contour model for image segmentation combining Cauchy-Schwartz distance and Gabor energy texture filters. We demonstrate the versatility of this approach on images from the Brodatz dataset and phase-contrast microscopy images of cells.

Keywords: Geometric active contours · Cauchy-Schwartz distance · Gabor energy · Texture feature segmentation

1 Introduction

Due to high throughput technology, a great influx of imaging data has become available in biomedical research producing large datasets that need to be processed in a reliable and unbiased way. As a result, there is an increased need for computer automation throughout the imaging framework [1] and in particular in the extension from high throughput to assays that include dynamic behaviour over time [2]. Existing image analysis frameworks are focused either on pre-processing the image to remove artifacts and enhance signal-to-noise ratio [3]; or using local intensity and texture information to delineate the cell surface from the background [4]. The latter category is non technology-specific and coupled with the ability to estimate parameters from data has the potential to unify detection techniques [5].

© Springer International Publishing AG 2017
A. Bracciali et al. (Eds.): CIBB 2016, LNBI 10477, pp. 12–26, 2017.
DOI: 10.1007/978-3-319-67834-4_2

Image segmentation is the task of partitioning an image into meaningful regions delineating objects and the background. Region-based segmentation takes into account the statistical properties of the image for example through density estimation techniques. Often the object regions are not Gaussian-distributed in pixel intensity making the detection by standard image analysis techniques (thresholding, edge-detection, region-based and connectivity preserving techniques) extremely challenging. This is the case in phase-contrast microscopy which is a widely used imaging technology, however images produced have low signal-to-noise ratio and illumination artifacts (bright halo around boundaries) caused by changes in object shape [3].

Active contour models are an unsupervised image segmentation technique consisting in defining a dynamic contour stretching over the object boundaries which partitions the image into distinct regions [6]. Geometric active contour models use an embedding of the contour into a higher dimensional surface (level set function) which is adapted to the information in the image until it converges to the object boundaries [7,8]. Geometric models overcome instability and topology problems of parametric active contours [6] and in addition enable probabilistic characterization of regions [9].

In this study, the Cauchy-Schwartz measure [10,11] of divergence is used to optimise image segmentation. Product-type measures such as Cauchy-Schwartz distance and Battacharyya distance [12] have numerical advantages over ratio-type measures including Kullback-Leibler [13] and Renyi's entropy in the approximation of region-specific distributions. By combining information theory, Gabor energy texture and a feature selection strategy, an automated segmentation strategy is described that can recover boundaries in textured images and challenging phase-contrast microscopy examples.

2   Materials and Methods

Let Ω_0 be a bounded open subset of $\mathbb{R}^2$ and let $I : \overline{\Omega}_0 \subset \mathbb{R}^2 \to \mathbb{R}$ represent an image. The partitioning of image Ω_0 into two non-overlapping regions: the target region Ω and the background region $\Omega_0 \setminus \Omega$ is defined by function $\mathbf{f} : \Omega_0 \subset \mathbb{R}^2 \to \mathbb{R}^n, \mathbf{f}(\mathbf{x}) = [f_1(\mathbf{x}), f_2(\mathbf{x}), ..., f_n(\mathbf{x})]^T$ which associates any image location $\mathbf{x} = (x, y) \in \mathbb{R}^2$ to a vector of features f_i. The dimension of the feature space is determined by the nature of features, e.g. $n = 1$ for grayscale intensity, $n = 3$ for color images or large n in the case of texture.

Features observed over the target and background regions represent random variables independently sampled from a target distribution, $p_t(\mathbf{f}(\mathbf{x}))$ and a background distribution $p_b(\mathbf{f}(\mathbf{x}))$ defined as:

$$p_t(\mathbf{f}(\mathbf{x})) = \frac{1}{||\Omega||} \int_\Omega K(\mathbf{f}(\mathbf{x}) - \mathbf{f}(\hat{\mathbf{x}}))d\hat{\mathbf{x}} \qquad (1)$$

$$p_b(\mathbf{f}(\mathbf{x})) = \frac{1}{||\Omega_0 \setminus \Omega||} \int_{\Omega_0 \setminus \Omega} K(\mathbf{f}(\mathbf{x}) - \mathbf{f}(\hat{\mathbf{x}}))d\hat{\mathbf{x}}$$

where $\hat{\mathbf{x}}$ denote uniformly distributed sampling locations from where the feature observations $\mathbf{f}(\hat{\mathbf{x}})$ are collected and $K(\mathbf{f}(\mathbf{x}))$ is a Parzen (Gaussian) density estimation kernel [14].

In the following, the use of the Cauchy-Schwartz information-theoretic measure is discussed as basis for defining a new image segmentation model. Cauchy-Schwartz distance is a measure of divergence between two distributions. It is part of a class of cross-entropy measures that includes Kullback-Leibler, Battacharyya and Renyi's entropy. The Cauchy-Schwartz distance is derived from the Cauchy-Schwartz inequality [10]:

$$||\mathbf{u}||^2||\mathbf{v}||^2 \le (\mathbf{u}^T\mathbf{v})^2 \Leftrightarrow -\log\frac{\mathbf{u}^T\mathbf{v}}{\sqrt{||\mathbf{u}||^2||\mathbf{v}||^2}} \ge 0. \tag{2}$$

where $\mathbf{u}$ and $\mathbf{v}$ are any two vectors. Cauchy-Schwartz distance is a product-type measure which alongside with Battacharyya distance has been shown to provide numerical advantages over ratio-type measures such as Kullback-Leibler in the approximation of region-specific distributions [11]. Given a partitioning of the image, region-specific p_t and p_b can be optimally estimated by modifying the partitioning in the direction of maximising Cauchy-Schwartz distance:

$$D_{CS}\left(p_t(\mathbf{f}(\mathbf{x})), p_b(\mathbf{f}(\mathbf{x}))\right) = -\log\frac{\int_{\mathbb{R}^n} p_t(\mathbf{f}(\mathbf{x}))p_b(\mathbf{f}(\mathbf{x}))d\mathbf{f}}{\sqrt{\int_{\mathbb{R}^n} p_b^2(\mathbf{f}(\mathbf{x}))d\mathbf{f}\int_{\mathbb{R}^n} p_t^2(\mathbf{f}(\mathbf{x}))d\mathbf{f}}} \ge 0. \tag{3}$$

2.1   Geometric Active Contour Model Based on Cauchy-Schwartz

The active contour partitioning of the image is represented using a level set function:

$$\Phi(\mathbf{x})\begin{cases} > 0, & \text{if } \mathbf{x} \in \Omega \\ < 0, & \text{if } \mathbf{x} \in \Omega_0 \setminus \Omega \\ = 0, & \text{if } \mathbf{x} \in \partial\Omega \end{cases}. \tag{4}$$

Maximising (3) is equivalent to minimising the argument of the logarithm. In the following, this is refered to as the Cauchy-Schwartz (CS) criterion:

$$CS(p_t(\mathbf{f}(\mathbf{x})), p_b(\mathbf{f}(\mathbf{x}))) = \frac{\int_{\mathbb{R}^n} p_t(\mathbf{f}(\mathbf{x}))p_b(\mathbf{f}(\mathbf{x}))d\mathbf{f}}{\sqrt{\int_{\mathbb{R}^n} p_t^2(\mathbf{f}(\mathbf{x}))d\mathbf{f}\int_{\mathbb{R}^n} p_b^2(\mathbf{f}(\mathbf{x}))d\mathbf{f}}}. \tag{5}$$

Let the notations be introduced:

$$G_1(\mathbf{x}, \Omega) = p_t(\mathbf{f}(\mathbf{x})); \ G_3(\mathbf{x}, \Omega) = \int_{R^n} p_t^2(\mathbf{f}(\mathbf{x}))d\mathbf{f}; \tag{6}$$

$$G_2(\mathbf{x}, \Omega) = p_b(\mathbf{f}(\mathbf{x})); \ G_4(\mathbf{x}, \Omega) = \int_{R^n} p_b^2(\mathbf{f}(\mathbf{x}))d\mathbf{f};$$

We define the Cauchy-Schwartz region-based geometric active contour model as:

$$J(\Phi) = \int_{\mathbb{R}^n} \frac{G_1(\mathbf{x}, \Omega)G_2(\mathbf{x}, \Omega)}{\sqrt{G_3(\mathbf{x}, \Omega)G_4(\mathbf{x}, \Omega)}}d\mathbf{f} + \mu\int_{\partial\Omega} ds = J_1(\Phi) + J_2(\Phi) \tag{7}$$

where $J_1(\Phi)$ is a region-based term enforcing the CS criterion (5) and $J_2(\Phi)$ is a boundary-based term enforcing minimum length of the contour. The evolution of $\Phi(\mathbf{f}(\mathbf{x}), t)$ from an initial given state $\Phi(\mathbf{f}(\mathbf{x}), 0) = \Phi_0(\mathbf{f}(\mathbf{x}))$ in the direction of minimising (7) is parameterised by $t \geq 0$.

We use shape derivation theory [15, 16] to obtain the Euler derivative of (7). The term $J_1(\Phi) = \int_{R^n} k(\mathbf{x}, \Omega) d\mathbf{f}$ is a region-based term with region-dependent descriptor $k(\mathbf{x}, \Omega) = G_1(\mathbf{x}, \Omega) G_2(\mathbf{x}, \Omega) G_3(\mathbf{x}, \Omega)^{-1/2} G_4(\mathbf{x}, \Omega)^{-1/2}$. Derivation leads to a summation of region-based terms with region-dependent descriptors detailed in the Appendix. Therefore, the Euler derivative of J_1 in the direction of $\mathbf{v}$ is:

$$dJ_{1r}(\Omega, \mathbf{v}) = \frac{A(\mathbf{x}, \Omega)}{||\Omega_0 \setminus \Omega||} \int_{\partial \Omega} \left(1 - \frac{G_2(\mathbf{x}, \Omega)}{G_4(\mathbf{x}, \Omega)}\right) G_1(\mathbf{x}, \Omega) * K(\mathbf{f}(\mathbf{x})) (\mathbf{v} \cdot \mathbf{n}) ds + \quad (8)$$
$$- \frac{A(\mathbf{x}, \Omega)}{||\Omega||} \int_{\partial \Omega} \left(1 - \frac{G_1(\mathbf{x}, \Omega)}{G_3(\mathbf{x}, \Omega)}\right) G_2(\mathbf{x}, \Omega) * K(\mathbf{f}(\mathbf{x})) (\mathbf{v} \cdot \mathbf{n}) ds$$

where $A(\mathbf{x}, \Omega) = G_3^{-1/2}(\mathbf{x}, \Omega) G_4^{-1/2}(\mathbf{x}, \Omega)$ and $*$ denotes convolution.

The term $J_2(\Phi)$ is a boundary-based term with boundary-independent descriptor, therefore $dJ_{2r}(\Omega, \mathbf{v}) = - \int_{\partial \Omega} \mu \mathrm{div} \left(\frac{|\nabla \Phi|}{||\nabla \Phi||}\right) (\mathbf{v} \cdot \mathbf{n}) ds$. The evolution equation becomes:

$$\frac{\partial \Phi}{\partial t} = \left[\frac{A(\mathbf{x}, \Omega)}{||\Omega||} \left(1 - \frac{G_1(\mathbf{x}, \Omega)}{G_3(\mathbf{x}, \Omega)}\right) (G_2(\mathbf{x}, \Omega) * K(\mathbf{f}(\mathbf{x}))) - \right. \quad (9)$$
$$\left. - \frac{A(\mathbf{x}, \Omega)}{||\Omega_0 \setminus \Omega||} \left(1 - \frac{G_2(\mathbf{x}, \Omega)}{G_4(\mathbf{x}, \Omega)}\right) (G_1(\mathbf{x}, \Omega) * K(\mathbf{f}(\mathbf{x}))) + \mu \mathrm{div} \left(\frac{|\nabla \Phi|}{||\nabla \Phi||}\right) \right] \mathbf{n}.$$

2.2 Gabor Energy Texture Features

Texture features include spatial information of pixel intensities. Commonly used in image processing is Gabor filtering which decomposes the image into sub-bands with a preferred orientation and spatial frequency by kernel convolution. The use of Gabor energy features sets the basis for a nonlinear multi-scale method of describing texture that resembles the way information is interpreted in the visual cortex [17, 18]. A 2D Gabor filter has the expression:

$$g_{\lambda, \sigma, \gamma, \theta, \varphi}(x, y) = e^{-\frac{x\prime^2 + \gamma^2 y\prime^2}{2\sigma^2}} \cos\left(2\pi \frac{x\prime}{\lambda} + \varphi\right)$$
$$x\prime = (x - x_0) \cos\theta + (y - y_0) \sin\theta \quad (10)$$
$$y\prime = -(x - x_0) \sin\theta + (y - y_0) \cos\theta$$

where $\theta \in [0 \ \pi)$ is the rotation angle of the gaussian envelope and λ and $\varphi \in (-\pi \ \pi]$ denote the spatial frequency and phase of the sinusoidal carrier. The Gaussian envelope is characterised by parameters γ, which specifies ellipticity and σ, a scaling parameter which controls the size of the Gaussian. The ratio σ/λ controls the number of parallel on and off stripes that the kernel contains.

This ratio is determined by the bandwidth b. In the following, we consider the case of $b = 1$ for which $\sigma = 0.56 \, \lambda$. The response of a Gabor filter (10) applied to an image is:

$$r_{\lambda,\sigma,\gamma,\theta,\varphi} = \int_{\Omega} I(u,v) g_{\lambda,\sigma,\gamma,\theta,\varphi}(x-u, y-v) du dv. \tag{11}$$

Gabor energy represents the combined magnitude of phase-shifted responses:

$$e_{\lambda,\sigma,\gamma,\theta}(x,y) = \sqrt{r^2_{\lambda,\sigma,\gamma,\theta,0}(x,y) + r^2_{\lambda,\sigma,\gamma,\theta,-\frac{\pi}{2}}(x,y)}. \tag{12}$$

Single Orientation Texture Features. Gabor energy feature function can be defined by discretising $\lambda = [\lambda_{min}, \lambda_{min} + \Delta\lambda, \dots]$, $\gamma = [\gamma_{min}, \gamma_{min} + \Delta\gamma, \dots]$ and $\theta = [\theta_1, \theta_2, \dots]$, $\theta_k = k\frac{\pi}{N}, k = \overline{0, N-1}$. Single orientation features are combined into a set:

$$\mathbf{f}^1 : \Omega_0 \in \mathbb{R}^n, \mathbf{f}^1(x,y) = [\mathbf{f}^1_{1,0}(x,y), \mathbf{f}^1_{1,1}(x,y) \dots \mathbf{f}^1_{n,N-1}(x,y)]^T \tag{13}$$

$$\text{where } \mathbf{f}^1_{n,k} = e_{\lambda_n,\gamma_n,\theta_k}(x,y).$$

Combined Orientation Texture Features. For textures without a preferred spatial orientation, combined Gabor energy features representing the superposition of Gabor energy terms for multiple orientations are added to a set:

$$\mathbf{f}^2 : \Omega_0 \in \mathbb{R}^n, \mathbf{f}^2(x,y) = [\mathbf{f}^2_1(x,y), \mathbf{f}^2_2(x,y) \dots \mathbf{f}^2_n(x,y)]^T \tag{14}$$

$$\text{where } \mathbf{f}^2_n(x,y) = \sum_{k=1}^{N} e_{\lambda_n,\gamma_n,\theta_k}(x,y).$$

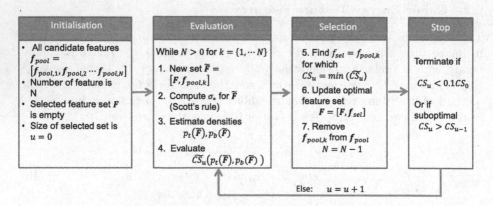

Fig. 1. Feature selection strategy produces a CS optimal feature set.

2.3 Feature Selection Strategy

The number of texture features increases computational complexity. This can be prevented by using a suitable feature selection strategy. In the following, a

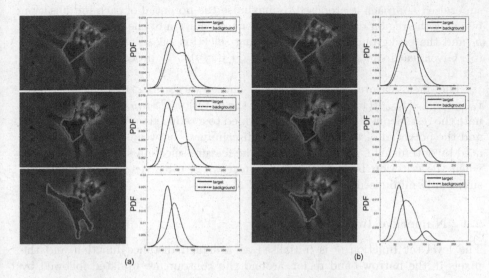

Fig. 2. Grayscale intensity segmentation using (a) Cauchy-Schwartz and (b) Kullback-Leibler models applied to a phase-contrast microscopy image of a cell: initialisation (top panel) is identical and evolution of the active contour is shown at intermediate (middle panel) and final (bottom panel) iterations accompanied by corresponding target and background distributions. Parameters $\mu = 0.001; w = 10$.

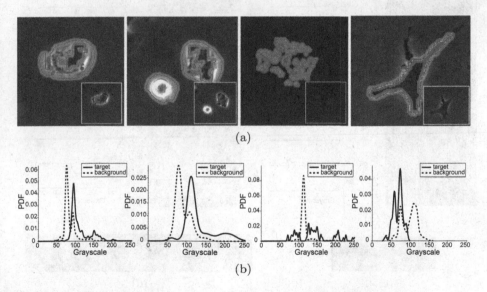

Fig. 3. Grayscale intensity segmentation using Cauchy-Schwartz model: (a) final iteration of active contour segmentation of phase-contrast microscopy images of cells (image inserts) with narrow band shown as a shaded region; (b) target and background distributions corresponding to contours in (a). Parameters $\mu = 0.2; w = 5$.

supervised approach is proposed to maximise the potential of each initialisation by using the region defined by Φ_0 as prior information of the densities of target p_t and the background p_b and performing selection based on the CS criterion (5). Given features $\mathbf{f}_{pool} = [\mathbf{f}_{pool,1}(x, y), \mathbf{f}_{pool,2}(x, y), \ldots]$, procedure is detailed in Fig. 1.

The first feature, $\mathbf{f}_{pool,u,1}(x, y)$, is chosen according to the procedure: $\mathbf{f}_{sel,0}(x, y) = \mathbf{f}_{pool,1}(x, y)$ and the value of the criterion $CS_0(p_t, p_b)$ is used to evaluate the rest. Sequentially, features are added to a reduced feature set $\mathbf{F} = [\mathbf{f}_{sel,u}]$ and the potential of the selected set to discriminate between p_t and p_b is evaluated by optimising $CS_u(p_t(\mathbf{F}), p_b(\mathbf{F}))$. The feature selection strategy terminates when the criterion becomes worse $CS_u(p_t, p_b) > CS_{u-1}(p_t, p_b)$ or when it is sufficiently minimised $CS_u(p_t, p_b) < 0.1 CS_0(p_t, p_b)$.

2.4   Numerical Implementation

The level set function Φ is initialised as a signed distance function and the pixels in the narrow-band region around the contour are updated followed by

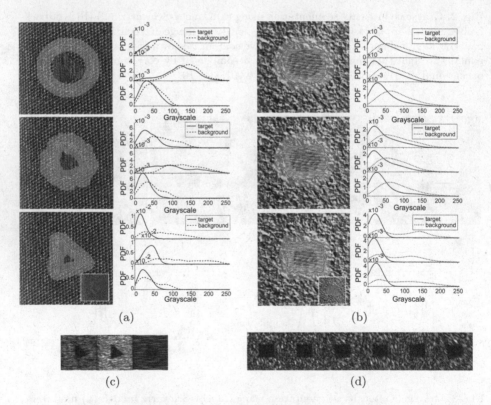

(a)   (b)

(c)   (d)

Fig. 4. Brodatz texture segmentation examples in images generated as fusion of two textures: (a, b) active contours evolving from initialisation (top), intermediate (middle) and final iteration (bottom) with corresponding estimated target and background distributions for three dominant features; (c) and (d) optimal feature sets corresponding to (a) and (b) respectively. Parameters $\mu = 0.2$, $w = 15$.

reinitialisation of the distance function to prevent numerical errors using the procedure in [19]. Density estimation was implemented using a (Parzen) Gaussian kernel with optimal variance obtained using Scott's rule [14] where n and m represent number of features and pixels respectively: $\sigma_{\mathbf{X}}^2 = \frac{1}{n} \sum_{i=1}^{n} \sigma_{ii}^2; \sigma^\star = \sigma_{\mathbf{X}} m^{\frac{1}{n+4}}$.

The narrow band technique was used to reduce computational complexity from $O(n^2)$ to $O(nk)$ where n and k represent the size of the image and of the narrow band region respectively [20]. Convergence was assesed from stationarity of the contour, i.e. less than 10% of pixels in the narrow band change sign in subsequent iterations. The geometric active contour parameters stiffness $\mu \in [0, 1]$ and width of the narrow band w are reported for each example.

3   Results

To demonstrate the ability of the Cauchy-Schwartz model to recover boundaries of objects, segmentation examples using grayscale are compared with an existing information theory-based active contour and limitations of using grayscale in phase-contrast microscopy images is discussed. Following this, Gabor energy texture segmentation is demonstrated on a number of Brodatz texture samples and phase-constrast microscopy images.

3.1   Segmentation of Phase-Contrast Images Based on Grayscale Intensity only Partially Recovers Boundaries

The CS-based geometric active contour was evaluated on images of cells acquired with a phase-contrast microscope (Fig. 2). Boundaries of the cell could be recovered in challenging examples where distributions of target and background regions showed significant overlap (Fig. 2a). We compared these results against a Kullback-Leibler (KL) active contour described in [13]. The KL model lead to faster convergence (Fig. 2b, 4 iterations) compared to CS (Fig. 2a, 11 iterations). Notably, cell debris was correctly excluded in the final contour by both models. However, by avoiding the local minimum in divergence visible in the final step of KL segmentation (Fig. 2b), the CS model recovered more of the object interior at the cost of increased number of iterations.

When applied to a wider phase-contrast microscopy image set results appeared mixed (Fig. 3) and boundaries were only partly recovered by the CS model (Fig. 3a) and no improvements were noted using KL (data not shown). Images with halo artifacts and the inclusion of dark and bright objects which are characteristic of phase-contrast, appeared to increase errors in the detection of target distributions (Fig. 3b). Overall, these examples indicated that not all microscopy images could be segmented using grayscale intensity alone. Given that the CS criterion showed increased detection compared to KL, we further tested the CS model by including information hidden in the texture characteristics of target and background regions.

3.2 Gabor Features Enable Detection of Noisy Object Boundaries in Textured Images

To investigate the ability to recover boundaries using Gabor texture features, test images were generated by fusing samples from the Brodatz [21] dataset (Fig. 4). The fused textures have similar mean intensity and contain noise thus resembling properties of microscopy images. A single orientation feature space was generated using $b = 1$, $\lambda = [1/15, 1/30, 1/60, 1/120, 1/240]$, $\gamma = [0.2, 0.4, 0.6, 0.8, 1]$; this was reduced to an optimal feature set using the CS-based feature selection strategy and the active contour was able to successfully recover the boundaries by estimating the target and background distributions. More features are selected in the example (Fig. 4b) and they appear more similar to each other when compared to (Fig. 4c). This suggested that a sparse feature set may be preferable to a finely sampled one.

3.3 Cauchy-Schwartz Model Detects Cells in Phase-Contrast Images Using Gabor Features

The performance of the geometric active contour and feature selection strategy were tested on real microscopy images displaying cells with bright and dark cell interior (Figs. 5 and 6). The texture of cells has no preferred orientation, therefore the feature space was combined from features at 8 different orientations followed by reduction to an optimal feature set. The active contour could detect each cell separately (Fig. 5) as well as jointly (Fig. 6). As expected, initial CS level exceeded the optimal threshold (final CS_u) obtained by the feature selections strategy (Fig. 1) but consistently fell under at large iteration numbers in all examples (Figs. 5c, d and 6b; dashed lines indicate optimal threshold). Boundaries of the dark cell (Fig. 5a) were easiest to detect as indicated by a large drop followed by approximately linear decay in the evolution of CS criterion (Fig. 5c). The bright cell example (Fig. 5b) posed increased difficulty in detection thus requiring larger iteration numbers compared to the dark one. In this case, the trend of the CS criterion showed a region of local minima followed by slow exponential decay (Fig. 5d). The combined bright and dark cell segmentation (Fig. 6a) proved the most challenging with multiple local minima and requiring the most iterations to achieve minimisation below the optimal threshold (Fig. 6b). These examples highlight that the problem of simultaneous segmentation of multiple objects with different intensity characteristics has an unexpectedly high level of difficulty when compared to the detection of individual objects. Nevertheless, the proposed feature selection and CS-based segmentation strategy is flexible enough to deal with either case and thus provides a solid basis for multiple object detection in microscopy images.

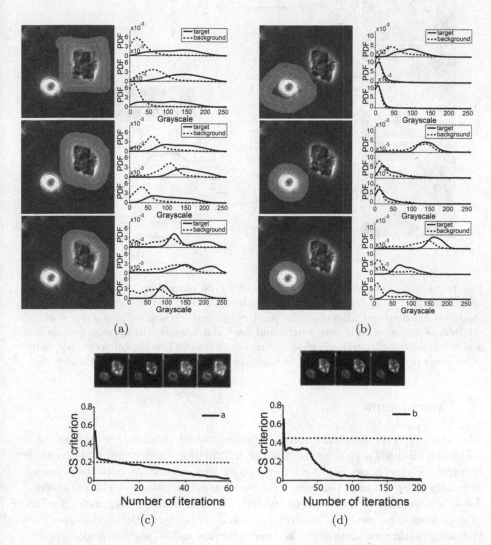

Fig. 5. Single cell detection using texture: (a, b) active contours evolving from initialisation (top), intermediate (middle) and final (bottom) interations applied to a cell with (a) dark and (b) bright cell interior; (c) and (d) show optimal features and criterion minimisation corresponding to (a) and (b) respectively. Parameters $\mu = 0.2; w = 15$.

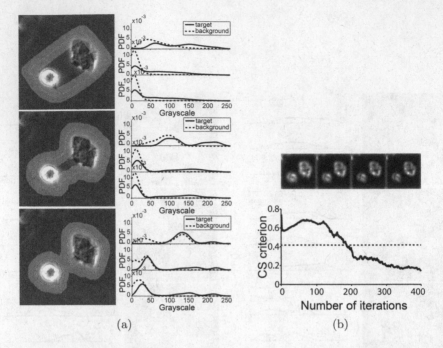

(a)   (b)

Fig. 6. Multiple cell detection using texture shows natural splitting of the contour to recover individual target regions: (a) active contours evolving from initialisation (top), intermediate (middle) and final (bottom) interations applied to a phase-contrast microscopy image of cells with bright and dark cell interior; (b) optimal feature sets and minimisation of criterion corresponding to (a); dashed line indicates optimal values of criterion predicted by the feature selection strategy. Parameters $\mu = 0.2; w = 15$.

4   Conclusions

The challenges of segmentation in phase-contrast microscopy images were addressed through a strategy combining information theory and Gabor energy features. A new image segmentation model was defined to optimise Cauchy-Schwartz (CS) distance between a desired (target) region and the background using a geometric active contour model. The CS model incorporates the use of a product-type measure of divergence and shape derivation techniques contributing to improved numerical accuracy. Similar to CS, segmentation based on Battacharyya distance was shown to improve detection compared to ratio-type measures [12]. Indeed in grayscale only segmentation, the CS model produced better separation between target and background regions in phase-contrast image of cells when compared to a Kullback-Leibler (KL) model [13] but at the cost of lower convergence speed. However, these results were confined to a subset of images exhibiting relatively smooth dark cell interior and boundaries of cells with mixed bright and dark appearance and halo artifacts failed to be detected by either CS or KL.

Texture information based on Gabor energy critically improved the recovery of boundaries of objects (either artificially generated or microscopic cells) with various intensity distributions. By incorporating Gabor energy features into the CS model, textured objects with geometric orientation could be recovered with single orientation features while microscopy images of cells which have non specific orientation required combined orientation texture features. The introduction of texture information posed the problem of increased computational complexity which was solved through a CS-specific feature selection strategy to ensure optimal segmentation.

Overall, this study introduces a unified approach to achieve active contour segmentation based on the Cauchy-Schwartz information theoretic measure. By the inclusion of unsupervised feature learning from training target and background datasets, this work could enable general detection of target objects without prior information of intensity distribution characteristics of the image. By extension to tracking it could address the lack of a generic platform for detection of multiple regions in biological images which is a major setback in the automation of high throughput analysis including dynamic behaviour over time.

Acknowledgments. VB was funded by a doctoral scholarship from The University of Sheffield. Authors kindly thank members of the Peter W Andrews Laboratory at the Centre for Stem Cell Biology for providing the microscopy images of cells.

Appendix

The derivative of J_1 in the direction of $\mathbf{v}$ is computed as:

$$dJ_{1r}(\Omega, \mathbf{v}) = \int_{R^n} k_s(\mathbf{x}, \Omega, \mathbf{v}) d\mathbf{f} - \int_{\partial\Omega} k(\mathbf{x}, \Omega)(\mathbf{v} \cdot \mathbf{n}) ds \qquad (15)$$

The shape derivative of k in the direction of $\mathbf{v}$ is given by:

$$k_s(\Omega, \mathbf{v}) = \frac{\partial k}{\partial G_1} dG_{1r}(\Omega, \mathbf{v}) + \frac{\partial k}{\partial G_2} dG_{2r}(\Omega, \mathbf{v}) + \frac{\partial k}{\partial G_3} dG_{3r}(\Omega, \mathbf{v}) + \frac{\partial k}{\partial G_4} dG_{4r}(\Omega, \mathbf{v}).$$
$$(16)$$

The term $G1$ is a region-based term with region-dependent descriptor.

$$G_1(\mathbf{x}, \Omega) = \int_{\Omega} H_1(\mathbf{x}, \Omega) d\hat{\mathbf{x}}; \quad H_1(\mathbf{x}, \Omega) = \frac{K(\mathbf{f}(\mathbf{x}) - \mathbf{f}(\hat{\mathbf{x}}))}{K_{11}(\mathbf{x}, \Omega)}; \qquad (17)$$

$$K_{11}(\mathbf{x}, \Omega) = \int_{\Omega} L_{11}(\mathbf{x}, \Omega) d\hat{\mathbf{x}}; \quad L_{11}(\mathbf{x}, \Omega) = 1.$$

The Eulerian derivative of $G1$ is:

$$dG_{1r}(\mathbf{x}, \Omega, \mathbf{v}) = \frac{1}{||\Omega||} \int_{\partial\Omega} \left(p_t(\mathbf{f}(\mathbf{x})) - K(\mathbf{f}(\mathbf{x}) - \mathbf{f}(s)) \right) (\mathbf{v} \cdot \mathbf{n}) \, ds. \qquad (18)$$

Similarly, the derivative of G_2 is:

$$dG_{2r}(\mathbf{x}, \Omega_0 \setminus \Omega, \mathbf{v}) = -\frac{1}{||\Omega_0 \setminus \Omega||} \int_{\partial \Omega} (p_b(\mathbf{f}(\mathbf{x})) - K(\mathbf{f}(\mathbf{x}) - \mathbf{f}(s))) \, (\mathbf{v} \cdot \mathbf{n}) \, ds.$$

(19)

Note that the expression (19) has a change of sign due to the normal vector $\mathbf{n}$ that changes direction w.r.t. the target and background region.

The term G_3 is a function of region-based terms:

$$G_3(\mathbf{x}, \Omega, \mathbf{v}) = \int_{R^n} H_3(\mathbf{x}, \Omega) df; \qquad H_3(\mathbf{x}, \Omega) = \frac{K_{31}(\mathbf{x}, \Omega)^2}{K_{32}(\mathbf{x}, \Omega)^2}.$$

(20)

The shape derivative of H_3 in the direction $\mathbf{v}$ of (20) is:

$$H_{3s}(\Omega, \mathbf{v}) = \frac{\partial H_3}{\partial K_{31}} dK_{31r} + \frac{\partial H_3}{\partial K_{32}} dK_{32r}$$

(21)

where the terms K_{31}, K_{32} are region-dependent terms with region-dependent descriptors and factorise as:

$$K_{31}(\mathbf{x}, \Omega) = \int_{\Omega} L_{31}(\mathbf{x}, \Omega) d\hat{\mathbf{x}}; \quad L_{31}(\mathbf{x}, \Omega) = K(\dot{\mathbf{f}}(\mathbf{x}) - \mathbf{f}(\hat{\mathbf{x}}));$$

(22)

$$K_{32}(\mathbf{x}, \Omega) = \int_{\Omega} L_{32}(\mathbf{x}, \Omega) d\hat{\mathbf{x}}; \quad L_{32}(\mathbf{x}, \Omega) = 1.$$

The corresponding derivatives are:

$$dK_{31r} = -\int_{\partial \Omega} K(\mathbf{f}(\mathbf{x}) - \mathbf{f}(\hat{\mathbf{x}})) \, (\mathbf{v} \cdot \mathbf{n}) \, ds$$

(23)

$$dK_{32r} = -\int_{\partial \Omega} (\mathbf{v} \cdot \mathbf{n}) \, ds.$$

Substituting (23) into (21), the expression for the derivative of G_3 becomes:

$$dG_{3r}(\mathbf{x}, \Omega, \mathbf{v}) = \frac{2}{||\Omega||} \int_{\partial \Omega} (G_3(\mathbf{x}, \Omega) - K(\mathbf{f}(\mathbf{x}) - \mathbf{f}(\hat{\mathbf{x}}))) \, (\mathbf{v} \cdot \mathbf{n}) \, ds.$$

(24)

Similarly, the derivation of the term G_4 with a sign change corresponding to the orientation of $\mathbf{n}$ w.r.t. the background region has the expression:

$$dG_{4r}(\mathbf{x}, \Omega_0 \setminus \Omega, \mathbf{v}) = -\frac{2}{||\Omega_0 \setminus \Omega||} \int_{\partial \Omega} (G_4(\mathbf{x}, \Omega) - K(\mathbf{f}(\mathbf{x}) - \mathbf{f}(\hat{\mathbf{x}}))) \, (\mathbf{v} \cdot \mathbf{n}) \, ds.$$

(25)

Substituting results (18), (19), (24), (25) into (16), the shape derivative of the descriptor k is obtained:

$$k_s(\Omega, \mathbf{v}) = \frac{1}{||\Omega||} \int_{\partial \Omega} G_2 G_3^{-1/2} G_4^{-1/2} (G_1 - K) \, (\mathbf{v} \cdot \mathbf{n}) \, ds$$

(26)

$$- \frac{1}{||\Omega_0 \setminus \Omega||} \int_{\partial \Omega} G_1 G_3^{-1/2} G_4^{-1/2} (G_2 - K) \, (\mathbf{v} \cdot \mathbf{n}) \, ds$$

$$- \frac{1}{||\Omega||} \int_{\partial \Omega} G_1 G_2 G_3^{-3/2} G_4^{-1/2} (G_3 - K) \, (\mathbf{v} \cdot \mathbf{n}) \, ds$$

$$+ \frac{1}{||\Omega_0 \setminus \Omega||} \int_{\partial \Omega} G_1 G_2 G_3^{-1/2} G_4^{-3/2} (G_4 - K) \, (\mathbf{v} \cdot \mathbf{n}) \, ds.$$

Following cancellation of duplicate terms and reordering, the final expression for the Eulerian derivative of criterion term J_1 is obtained:

$$
\begin{aligned}
dJ_{1r}(\Omega, \mathbf{v}) = & -\frac{A(\mathbf{x}, \Omega)}{\|\Omega\|} \int_{\partial \Omega} \left(1 - \frac{G_1(\mathbf{x}, \Omega)}{G_3(\mathbf{x}, \Omega)} \right) \\
& \int_{\mathbb{R}^n} G_2(\mathbf{x}, \Omega) K(\mathbf{f}(\mathbf{x}) - \mathbf{f}(\hat{\mathbf{x}})) d\mathbf{f}(\mathbf{v} \cdot \mathbf{n}) \, ds \\
& + \frac{A(\mathbf{x}, \Omega)}{\|\Omega_0 \setminus \Omega\|} \int_{\partial \Omega} \left(1 - \frac{G_2(\mathbf{x}, \Omega)}{G_4(\mathbf{x}, \Omega)} \right) \\
& \int_{\mathbb{R}^n} G_1(\mathbf{x}, \Omega) K(\mathbf{f}(\mathbf{x}) - \mathbf{f}(\hat{\mathbf{x}})) d\mathbf{f}(\mathbf{v} \cdot \mathbf{n}) \, ds
\end{aligned}
\tag{27}
$$

where $A(\mathbf{x}, \Omega) = G_3^{-1/2}(\mathbf{x}, \Omega) G_4^{-1/2}(\mathbf{x}, \Omega)$.

References

1. Eliceiri, K.W., Berthold, M.R., Goldberg, I.G., et al.: Biological imaging software tools. Nat. Methods **9**(7), 697–710 (2012)
2. Allison, T.F., Powles-Glover, N.S., Biga, V., Andrews, P.W., Barbaric, I.: Human pluripotent stem cells as tools for high-throughput and high-content screening in drug discovery. Int. J. High Throughput Screen. **5**, 1–13 (2015)
3. Yin, Z., Kanade, T., Xu, D., Fisher, J.: Understanding the phase contrast optics to restore artifact-free microscopy images for segmentation. Med. Image Anal. **16**(5), 1047–1062 (2012)
4. Dewan, M.A., Ahmad, M.O., Swamy, M.N.: A method for automatic segmentation of nuclei in phase-contrast images based on intensity, convexity and texture. IEEE Trans. Biomed. Circuits Syst. **8**(5), 716–728 (2014)
5. Held, C., Palmisano, R., Haberle, L., et al.: Comparison of parameter-adapted segmentation methods for fluorescence micrographs. Cytometry A **79**(11), 933–945 (2011)
6. Blake, A., Zisserman, A.: Visual Reconstruction. MIT Press, Cambridge (1987)
7. Caselles, V., Catte, F., Coll, T., et al.: A geometric model for active contours in image processing. Numerische Mathematik **66**, 1–31 (1993)
8. Malladi, R., Sethian, J.A., Vemuri, B.C.: Shape modeling with front propagation: a level set approach. IEEE Trans. Pattern Anal. Mach. Intell. **17**, 158–175 (1995)
9. Cremers, D., Rousson, M., Deriche, R.: A review of statistical approaches to level set segmentation. Int. J. Comput. Vis. **72**, 195–215 (2007)
10. Principe, J., Xu, D., Fisher, J.: information theoretic learning. In: Unsupervised Adaptive Filtering. Wiley, New York (2000)
11. Gokcay, E., Principe, J.C.: Information theoretic clustering. IEEE Trans. Pattern Anal. Mach. Intell. **24**(2), 158–171 (2002)
12. Michailovich, O.V., Rathi, Y., Tannenbaum, A.: Image segmentation using active contours driven by the Battacharya gradient flow. IEEE Trans. Image Process. **16**(11), 2787–2801 (2007)
13. Houhou, N., Thiran, J.P., Bresson, X.: Fast texture segmentation model based on the shape operator and active contour. Proc. IEEE Comput. Vis. Pattern Recognit. **53**, 1–8 (2008)

14. Scott, D.W.: Multivariate Density Estimation: Theory, Practice and Visualisation. Probability and Statistics. Wiley, New York (1992)
15. Jehan-Besson, S., Barlaud, M., Aubert, G.: Video object segmentation using Eulerian region-based active contours. In: International Conference on Computer Vision, pp. 353–361 (2001)
16. Aubert, G., Barlaud, M., Faugeras, O., et al.: Image segmentation using active contours: calculus of variations or shape gradients. SIAM J. Appl. Math. **63**, 2128–2154 (2002)
17. Petkov, N., Kruizinga, P.: Computational models of visual neurons specialised in the detection of periodic and aperiodic visual stimuli: bar and grating cells. Biol. Cybern. **76**(2), 83–96 (1997)
18. Kruizinga, P., Petkov, N.: Nonlinear operator for oriented texture. IEEE Trans. Image Process. **8**, 1395–1407 (1999)
19. Sussman, M., Smerenca, P., Osher, S.J.: A level set approach for computing solutions to incompressible two-phase flow. J. Comput. Phys. **114**, 146–159 (1994)
20. Adalsteinsson, D., Sethian, J.: A fast level set method for propagating interfaces. J. Comput. Phys. **118**, 269–277 (1995)
21. http://www.ux.uis.no/~tranden/brodatz.html

Host Phenotype Prediction from Differentially Abundant Microbes Using RoDEO

Anna Paola Carrieri[1], Niina Haiminen[2], and Laxmi Parida[2]([✉])

[1] IBM Research UK, Warrington WA4 4AD, UK
[2] IBM T.J. Watson Research Center, Yorktown Heights, NY 10598, USA
parida@us.ibm.com

Abstract. Metagenomics is the study of metagenomes which are mixtures of genetic material from several organisms. Metagenomic sequencing is increasingly used in human and animal health, food safety, and environmental studies. In these high-dimensional (metagenomic) data, the phenotype of the host organism, e.g., human, may not be obvious to detect and then the ability to predict it becomes a powerful analytic tool. For example, consider predicting the disease status of an individual from their gut microbiome.

In this study, we compare various normalization methods for metagenomic count data and their impact on phenotype prediction. The methods include RoDEO, Robust Differential Expression Operator, originally developed for gene expression studies. The best prediction accuracy is observed for RoDEO-processed count data with linear kernel support vector machines in most cases, for a variety of real datasets including human, mouse, and environmental samples.

We also address the problem of identifying the most relevant microbial features that could give insight into the structure and function of the differential communities observed between phenotypes. Interestingly, we obtain similar or better phenotype prediction accuracy with a small subset of features as with the complete set of sequenced features.

Keywords: Metagenomics · Phenotype prediction · Differential abundance · Feature selection

1 Scientific Background

Technological advances in high-throughput sequencing and annotation now allow the characterization of genomes, transcriptomes, and most recently metagenomes as part of everyday research in many fields. While single-gene, usually 16S ribosomal RNA (rRNA), sequencing can be used to infer bacterial community members, whole-genome shotgun sequencing can reveal details of the activity and function of the microbial community. Meta-transcriptomic sequencing can be applied to further investigate the actively transcribed sequences. One of the

---

A.P. Carrieri and N. Haiminen contributed equally to this work.

A. Bracciali et al. (Eds.): CIBB 2016, LNBI 10477, pp. 27–41, 2017.
DOI: 10.1007/978-3-319-67834-4_3

major research challenges of the current decade is gaining insight into the structure, organization, and function of microbial communities which will be enabled by both experimental and computational metagenomic analyses [1].

Since the sequencing methods yield relative rather than absolute gene or species counts, a fundamental methodological question of appropriate normalization and scaling of the counts arises. Approaches such as using the raw counts, log-transformed counts, length-normalized counts, and other normalization methods have been investigated [2–4]. We propose applying RoDEO projection as a pre-processing method for metagenomic counts.

RoDEO (Robust Differential Expression Operator, http://researcher.watson. ibm.com/group/5513) [5] was originally designed for detecting differentially expressed genes from single species RNA-sequencing data. The underlying nonparametric model and ranking-based ordering of genes can be applied in the context of various count data, including species counts from metagenomic samples. We apply RoDEO on metagenomic count data due to its robust design that does not rely on any assumptions regarding the underlying count distributions, and its applicability even in the absence of replicate samples, a common characteristic of metagenomic data.

In this paper we investigate the task of predicting the phenotype of the host organism (or environment) starting from OTU (Operational Taxonomical Unit, e.g., species or genus) counts. This question is relevant, for example, if we aim to predict the disease state from gut or fecal microbiome samples of humans and animals [7,8]. A recent related work on the topic includes a study of approaches to metagenomics-based prediction tasks and potential strength of microbiome-phenotype associations [9].

We investigate the effect of RoDEO projection on the prediction accuracy, and contrast it with existing normalization methods, namely Log-transformation, DESeq2 [10], and CSS (Cumulative Sum Scaling) [2]. We compare several kernel options for SVM (Support Vector Machine) prediction. SVMs are well established fundamental machine learning methods that have been applied in genomic, transcriptomic, and recently also in the microbial phenotype prediction context [11]. We find that the linear kernel SVM yields the best accuracy values across all the datasets and normalization methods. We also consider Random Forests (RF) [12] as they are state-of-the-art classification approaches and are appropriate for this type of data [13].

Furthermore, we investigate the problem of identifying a subset of OTUs that are important for differentiating the phenotypes. The process of selecting a subset of features consists of reducing the size of an high-dimensional dataset to retain only relevant, differentiating features [14]. We apply feature selection by identifying the most differentially abundant OTUs between the phenotype sample groups, and use them for predicting the host phenotype. The top differentiating OTUs are selected using two differential gene expression methods RoDEO DE and DESeq2.

We show that the prediction accuracy obtained selecting the top differential 20 OTUs is comparable, if not higher, to using the entire set of OTUs across all the datasets we consider in our experiments. Although RoDEO DE and DESeq2 yield different sets of top differentiating OTUs, the prediction accuracy values obtained using the different OTUs subsets are very close. While the prediction accuracy obtained using RF is often higher or comparable to the one obtained using SVM with linear kernel, RF is more resource consuming especially for a large number of features, i.e. when we use the entire set of OTUs for the prediction.

2   Materials and Methods

In this section we describe the various normalization, differential abundance, and phenotype prediction methods, as well as the datasets used in this study.

2.1   RoDEO Normalization

RoDEO sequence count data normalization, called *projection*, is not focused on the relative counts of reads for each OTU, but on the relative order of the counts within a sample. The count values of all OTUs in an experiment are utilized in a re-sampling approach, to determine robust relative ranks of the genes in several re-sampled instances of the sequencing experiments. A global parameter P determines the number of possible output values of the projection, ensuring that samples processed with the same P are comparable.

The *projection* process of RoDEO takes as input count data, such as the number of reads mapping to a OTU, and performs repeated re-sampling of the reads falling on the OTUs. In this way RoDEO projection process obtains a distribution which represents several randomized draws of sequencing reads from the input sample, according to the initial OTU abundances. In each re-sampling, the reads falling onto each OTU are counted, the OTUs are ranked by decreasing count, and the cumulative curve of the counts is optimally divided into segments $1, ..., P$. The number of segments P defines the resolution at which DE genes are discovered. We choose P for each dataset according to the number of (non-zero) entries per sample. In the RoDEO publication [5] we use 15–20 segments for human and plant data with tens of thousands of genes. Thus the dimensionality of the sequence count data is reduced from thousands of distinct values onto a small number of P possible values.

The projection and re-sampling makes RoDEO resilient in the presence of noisy and sparse count data with a large value range, such as observed in metagenomic sequencing data, and on a previous application on plant gene expression data [5].

2.2   DESeq2 Normalization

DESeq2 [10] is a well known method designed for differential analysis of count data using shrinkage estimation for sequence count dispersions. In a recent work which evaluates several methods for the identification of differentially abundant genes between metagenomes [4], DESeq2 was found to be among the best approaches for the task.

2.3 Other Normalization Methods

The baseline for comparing RoDEO to other methods of processing the counts is using the raw sequence counts per OTU. Log-transformation is a standard pre-processing step for sequence count data applied in many studies, including the respective studies for the datasets analysed in our paper [2, 7, 15]. Therefore we take the log of the count data (after adding 1 to all the counts we use the log function in R to compute the natural logarithm).

In addition, we evaluate prediction results on the CSS method as implemented in QIIME. CSS [2] was introduced in conjunction of the mouse microbiome dataset that is included in our study. According to the authors, CSS corrects the bias in the assessment of differential abundance. We include this method in our evaluation since it appears better than DESeq (previous version of DESeq2) for the class separation task studied on the mouse dataset.

2.4 RoDEO Differential Abundance

Differential Abundance of an OTU between two groups is computed as a DA score (analogously to differential expression, DE, in the gene expression context). This score takes into account the projected distributions for each sample in the two groups. In this work we use the *mean* distance between the projected distributions instead of mode used in the original paper. The final score for an OTU is the mean distance between the phenotype group projected distributions for this OTU multiplied by the max. norm distance (measuring overlap) between the distributions.

In order to evaluate datasets at different scales, with different numbers of non-zero OTU counts and total counts, we apply *scaling* [6]. The main idea is, we use a different value for the number of projected values P, depending on the count distributions in the samples. Details on this process on the studied datasets are provided in the Appendix.

2.5 DESeq2 Differential Abundance

DESeq2 provides both a normalization function, and a DE score computation function; we use the resulting DE values as the DA per OTU, obtained from the QIIME [16] microbiome analysis pipeline (version 1.9.1).

2.6 Phenotype Prediction

Support Vector Machines (SVMs) are among the most powerful and versatile binary classifiers used in a myriad of applications. We evaluate SVMs with linear, polynomial, radial and sigmoid kernels for phenotype prediction on three different metagenomic datasets described in Sect. 2.7.

We conduct 10-fold Cross Validation (CV), repeating the process 100 times, on the four different trained SVM kernels on RoDEO projected counts, log-transformed counts, as well as the CSS and DESeq2 processed counts. We report the accuracy of each prediction as the percentage of correct phenotype calls for the test set and we include the Matthews Correlation Coefficient (MCC). The latter coefficient is a measure of the quality of binary classification that can be used even when the two classes are of very different sizes. MCC can assume values between +1 and −1, where +1 indicates a perfect prediction, 0 no better than random and −1 represents total disagreement between predictions and observations.

After performing 10-fold CV process 100 times, we compute the average of the 100 accuracy and MCC values for each combination of kernel and dataset. The average accuracy and MCC values are summarised in Table 1, while the distribution of accuracy values and their average are visualized in Fig. 1.

Furthermore in Sect. 3.2, we apply, to the whole set of OTUs and to selected subsets of OTUs, SVM with linear kernel together with another prediction method, Random Forests, in order to compare their respective prediction accuracy, MCC and F1 score values.

The F1 measure is widely applied in information retrieval for measuring document classification. F1 score has an intuitive meaning: it tells how precise the classifier is (how many instances it classifies correctly), as well as how robust it is (it does not miss a significant number of instances). In statistical analysis of binary classification, the F1 score (which reaches its best value at 1 and worst at 0) is a measure of test accuracy and can be interpreted as a weighted average of the precision and recall.

The SVM and RF prediction is computed using the svm() and rf() R functions (e1071 package). All phenotype prediction results and figures have been produced using R (version 3.2.3).

2.7   Datasets

We investigate the accuracy of phenotype prediction starting from three different available metagenomic datasets: human, mouse, and corpse decomposition data. The human metagenome sequences originate from genome-wide shotgun sequencing, while the mouse and corpse data result from targeted rRNA sequencing. We obtained directly the read counts per OTU in each sample. For more details on the datasets please see the original publications.

Human dataset [7] consists of fecal metagenome of 70-year-old European women with either Normal Glucose Tolerance (NGT) or Type 2 Diabetes (T2D). Though T2D is caused by a complex combination of lifestyle and genetic factors, an altered gut microbiome has been linked to metabolic diseases including obesity, diabetes and cardiovascular disease. All microbiome samples were sequenced with Illumina HiSeq2000, and aligned to 2,382 non-redundant reference genomes (from the National Center for Biotechnology Information (NCBI) and Human Microbiome Project (HMP databases) in order to determine the composition of the gut microbiota. In our study we consider 43 NGT and 53 T2D samples described by a total of 134 OTUs at the family level. The phenotypes for the human dataset are healthy (NGT, 43 samples) and sick (T2D, 53 samples).

Mouse microbiome data [2] consist of mice fecal samples. Mice were fed with either Western (W) or Low-Fat, Plant Polysaccharide-rich (LF-PP) diet. Fecal samples for each mouse went through Polymerase Chain Reaction PCR amplification of the bacterial 16S rRNA gene V2 region. OTUs were classified by RDP11 and annotated. We analyze the dataset composed of 139 samples and 10,172 OTUs. The phenotypes for this dataset are W diet (54 samples) and LF-PP diet (85 samples).

Corpse microbiome data [15] consist of time-series samples from donated human bodies exposed to all natural elements. Two corpses were placed during the spring for 82 days and two corpses were placed during the winter for 143 days. Samples from multiple skin and soil locations were taken at different time points, daily or every other day the first month and less frequently thereafter. 16S rRNA gene (archaeal and bacterial community), 18S rRNA gene (microbial eukaryotic community), and ITS region (fungal community) were sequenced with high-throughput amplicon-based sequencing technology to characterize the full microbial diversity associated with decomposition. Sequence reads were classified into OTUs on the basis of sequence similarity using QIIME. We examine the read counts of 213 samples, having sum of counts above 10, taken from the left knee (skin and soil) at all the time points. There are a total of 17,803 OTUs observed in these samples. We choose this particular body site as it is sampled for both spring and winter conditions with sufficient detail, and there are many non-zero OTUs shared between the two conditions. The phenotypes for the corpse dataset are spring (79 samples) and winter (134 samples).

3   Results

In this section we summarize the phenotype prediction results on full datasets and on selected top differentially abundant features.

3.1   Phenotype Prediction on Full Datasets

Figure 1 summarizes visually the average prediction accuracy for each dataset and kernel, while Table 1 shows in more detail the differences in average prediction accuracy and MCC across the methods and highlights the best results per dataset. The results show that average accuracy and MCC consistently indicate the same combination of normalization and kernel as best for a particular dataset.

Human dataset has the lowest prediction accuracy and the lowest Matthews correlation coefficient. On this data RoDEO is best for nearly every kernel, and especially clearly improves the linear kernel prediction, yielding the best overall accuracy of 67.38% and the best MCC of 0.34.

The mouse data prediction is nearly perfect for most kernels and normalization methods. Only the Log data with sigmoid and radial kernels, as well as DESeq2 and CSS with polynomial kernel have lower accuracy.

On the corpse data, different kernels have quite different behavior. The worst seems to be sigmoid kernel and again the best is the linear kernel, where CSS slightly improves over RoDEO and yields 96.3% accuracy and 0.92 MCC, compared to 96.0% accuracy and 0.91 MCC of RoDEO.

Human prediction accuracy is not as high as for the other datasets studied here; in the original study they improve it by assembling novel entities from the unmapped reads and using them as additional features for prediction. This demonstrates there is still significant relevant content in the microbiomes that have not been encountered and annotated before. Still, in the mouse and corpse datasets using sequences mapped against existing databases yield highly accurate separation of phenotypes.

Most importantly, the best prediction accuracy is observed for RoDEO processed data in most cases and for the linear kernel. CSS is the second best method, followed by DESeq2 and Log. Also note that RoDEO clearly improves prediction accuracy on the clinically relevant human dataset, improving the chances of correctly diagnosing Type 2 Diabetes based on the gut microbiome.

Table 1. Accuracy as the average percentage of correct phenotype predictions in the cross validation results using linear, polynomial, radial, and sigmoid kernels. The values in the accuracy table correspond to the rightmost plots in Fig. 1. On the right, Matthews correlation coefficient (MCC) values are reported for each dataset and method. The best accuracy and MCC values are reported in black bold text.

		Accuracy (%)				MCC			
		Lin	Pol	Rad	Sig	Lin	Pol	Rad	Sig
Human	RoDEO	**67.38**	67.00	62.40	55.72	**0.34**	0.33	0.26	0.00
	Log	56.38	55.25	63.08	56.70	0.12	0.10	0.24	0.03
	DESeq2	56.00	57.60	63.00	55.71	0.12	0.15	0.24	0.0
	CSS	58.40	60.31	55.70	55.71	0.17	0.20	0.06	0.0
Mouse	RoDEO	**100.0**	99.97	**100.0**	98.55	**0.999**	0.998	**0.999**	0.968
	Log	99.99	99.90	76.64	61.86	0.998	0.997	0.514	0.087
	DESeq2	**100.0**	61.15	**100.0**	99.99	**0.999**	0.0	**0.999**	0.998
	CSS	**100.0**	94.11	**100.0**	99.98	**0.999**	0.883	**0.999**	0.998
Corpse	RoDEO	96.0	93.90	94.47	49.33	0.91	0.86	0.88	−0.01
	Log	82.7	75.75	62.9	56.87	0.63	0.51	0.0	0.01
	DESeq2	94.8	83.4	93.7	65.56	0.88	0.65	0.87	0.27
	CSS	**96.3**	81.27	93.6	93.7	**0.92**	0.60	0.86	0.86

3.2   Phenotype Prediction on Selected Features

In order to establish a baseline on the de-duplicated datasets we use for feature selection, as discussed in the Appendix, we first evaluate prediction accuracy

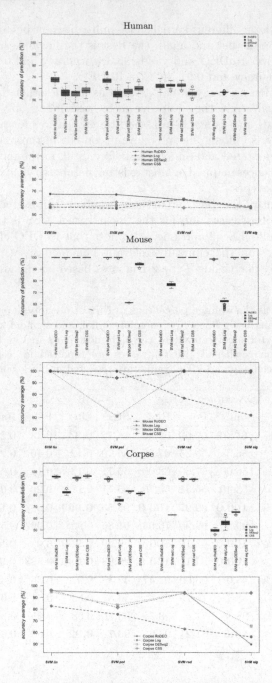

Fig. 1. Phenotype prediction accuracy in the 100 iterations of 10-fold Cross Validation for each SVM kernel (linear, polynomial, radial, sigmoid), RoDEO processed data and other normalization methods for human, mouse, and corpse data. For each dataset, the distribution plot of average accuracy across 100 iterations is followed by the corresponding overall average accuracy plot.

using all the OTUs. We compute 10-Fold Cross Validation using SVM with linear kernel (as we show in Sect. 3.1 that the best prediction accuracy overall is obtained with linear kernel) and Random Forest prediction methods. The average accuracy, MCC and F1 score values obtained for each dataset and each normalization and prediction method are shown in Table 2 in the "All OTU" columns. For the mouse dataset, similarly to the results shown in Table 1, accuracy is near perfect for all prediction methods, thus omitted from this evaluation.

Next we apply RoDEO and DeSeq2 differential expression methods to RoDEO projected data and DeSeq2 normalized data, respectively, and rank the OTUs according to their differential abundance (DA) scores for all three datasets. We select the top X where $X = 2, \ldots, 50$ most differential abundant OTUs and perform 10-fold CV on these subsets of different sizes using SVM linear and RF, to evaluate the prediction accuracy using the selected features only. The results are shown in Fig. 2. The horizontal lines denote the accuracy values reported in Table 2 for all OTUs. Using the most differentially abundant OTUs allows us to achieve similar or even better accuracy, MCC and F1 score compared to using the whole set of OTUs.

Based on the results in Fig. 2, we choose the value $X = 20$ as a representative small number of OTUs that yields phenotype prediction accuracy comparable

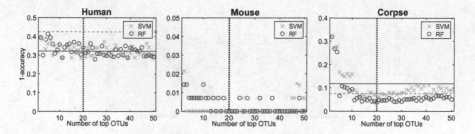

Fig. 2. RoDEO processed SVM and RF 10-fold CV results on varying numbers of top OTUs. Horizontal lines (SVM dashed, RF solid) denote the accuracy values when using all OTUs.

Table 2. Accuracy, MCC and F1 average values of 10 cross-fold validation results using linear kernel SVM and Random Forest prediction methods and considering either the top 20 DA OTUs or the complete set of OTUs. The best accuracy, MCC and F1 values for each dataset is shown in bold text.

		Accuracy (%)				MCC				F1			
		Subset 20		All OTU		Subset 20		All OTU		Subset 20		All OTU	
		RF	SVM	RF	SVM	RF	SVM	RF	SVM	RF	SVM	RF	SVM
Human	RoDEO	**70.11**	66.22	67.88	57.55	**0.42**	0.32	0.33	0.05	**0.71**	0.69	0.71	0.60
	DESeq2	65.66	58.88	61.00	59.66	0.32	0.19	0.22	0.15	0.68	0.58	0.65	0.61
Corpse	RoDEO	93.93	**94.39**	88.2	92.44	0.86	**0.88**	0.75	0.84	0.89	**0.92**	0.83	0.89
	DESeq2	93.42	89.67	86.47	93.85	0.86	0.79	0.71	0.86	0.90	0.85	0.79	0.89

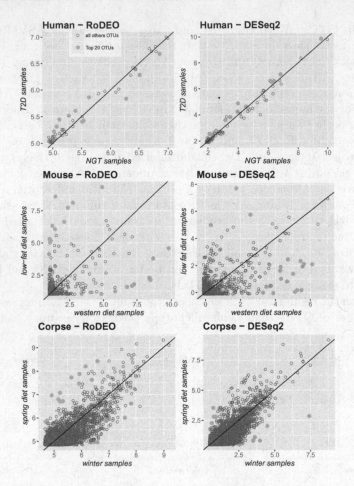

Fig. 3. Visualization of all OTUs (blue) and top 20 differentially abundant OTUs (red), for RoDEO (left) and DESeq2 (right) processed data. Each dot represents the average value of RoDEO projected or DESeq2 normalized samples having one phenotype (x) versus the other phenotype (y). The scale in each plot corresponds to either RoDEO projected values or DESeq2 normalized values. (Color figure online)

to all OTUs for all three datasets. In the following, we study in detail using the top 20 differentially abundant OTUs for phenotype prediction.

Table 2 shows that linear kernel SVM and RF methods using the whole set of OTUs or the top 20 OTUs give overall similar accuracy/MCC and F1 score results over all the three datasets. Furthermore, the results show that accuracy, MCC and F1 score are consistent as they indicate the same best combination of normalization, DA and kernel methods for a particular dataset. For the human dataset the best prediction result is given by RF method using RoDEO projected data and its subset of 20 top DA OTUs. For the corpse dataset the best prediction

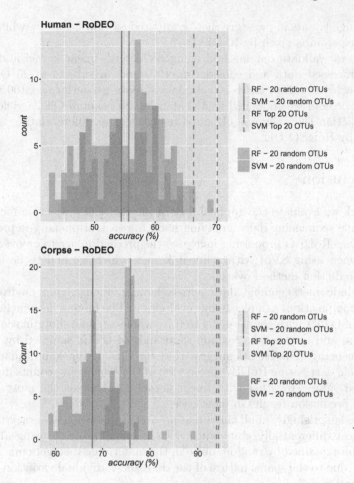

Fig. 4. The histogram shows the distribution of average phenotype prediction accuracy in 10-fold CV of 100 random subsets of 20 OTUs. Solid lines represent the average accuracy of the random OTUs subsets, while dashed lines show the average accuracy of 10-fold CV obtained using only the 20 top DA OTUs.

is obtained with linear kernel SVM on RoDEO projected data using the subset of 20 top DA OTUs.

The two methods, RoDEO DE and DeSeq2, yield different sets of top OTUs for all three datasets, but the prediction accuracy, MCC, and F1 scores on them are still quite close. Although the exact OTU names are different, representatives from the same family are selected by both methods, such as *Lachnospiraceae* for mouse.

Figure 3 shows details about the 20 OTUs in each dataset and for both RoDEO and DESeq2 methods. Overall the normalized datasets look quite similar between methods, but there are some differences, also regarding the values for the selected top OTUs. For example, DESeq2 appears to select many OTUs that have high counts in "western diet" compared to "low fat diet", while a more balanced selection is given by RoDEO.

Finally, we validate our method using SVM with linear kernel and RF for RoDEO processed data and considering 100 random subsets of 20 OTUs for each dataset. For each of these 100 random subsets we performed 100 times 10 CF validation and we show in Fig. 4 that using 20 random OTUs yield clearly worse prediction than the one obtained using the top differentiating 20 OTUs computed by RoDEO DE.

4   Conclusion

In this work we evaluate the applicability of the RoDEO projection method for metagenomic sequencing data, applying it on the task of phenotype prediction. We show that RoDEO processing increases the prediction accuracy over current methods when using SVM with a linear kernel, which we find to be the most accurate prediction method overall.

We include metagenomic data across human, mouse, and environmental (corpse decomposition) samples in our evaluation. The human data includes only a handful of OTUs with counts generated by whole-genome shotgun sequencing, while mouse and corpse data include thousands of OTUs sampled by targeted region sequencing. The results suggest that for various types and quantities of metagenomic data, using RoDEO projection of the sequencing counts onto lower dimensional values, together with linear kernel SVM yields the most accurate phenotype prediction results in most cases.

Perhaps surprisingly, in all three real datasets, prediction accuracy using the top few most differentially abundant OTUs is comparable to using all OTUs. This may be explained by random noise in the underlying metagenomic sequencing results, due to the sparse nature of the data and individual variation between the biological samples.

The actual top OTUs selected vary between the RoDEO and DESeq2 methods, but both provide accurate phenotype predictions using the respective OTUs. This indicates potential for accurate disease diagnostics and other phenotype prediction tasks by measuring a handful of most differential features only. RoDEO projection and feature selection, combined with either RF or SVM prediction yields consistently accurate phenotype prediction results.

Appendix: Experimental Details

RoDEO Projection Details on Full Datasets

For each of the 96 human samples with 134 OTUs, we run RoDEO for 100 independent re-sampling simulations, with $P = 7$ number of segments, 10^6 number of reads for the re-sampling and gap parameter equal to 1. For each of the samples we compute the average of projected values for each OTU (average of the 100 iterations), and combine all the obtained values in a single matrix.

Similarly, we apply RoDEO to the 139 mouse samples and 10,172 OTUs for 100 independent re-sampling simulations, with $P = 10$ number of segments, 10^7 number of reads for the re-sampling and gap equal to 1, and we compute the average of projected OTU values.

Finally, we run RoDEO for each of the 213 corpse samples with 17,803 OTUs for 100 independent re-sampling simulations, with $P = 10$ number of segments, 10^7 number of reads for the re-sampling and gap between the samples equal to 2. In the same way as described before, we compute the average of projected OTU values for each sample.

Feature Selection Details

We start the feature selection process deleting duplicated OTUs from each of the three initial raw count datasets described in Sect. 2.7. Removing identical OTUs allow us to deal with smaller datasets and apply Random Forests as an alternative prediction method to SVM. More precisely, for the corpse data we remove about 3000 OTUs passing from an original dataset of 213 samples and 17804 OTUs to a new dataset with 213 samples and 14789 OTUs. For the mouse data we pass from 139 samples described by 10172 OTUs to 139 samples described by only 4411 features. Finally, in the human data we find only 4 OTUs identical in the count and we obtain a new human dataset with 97 samples and 130 OTUs.

We proceed to run DESeq2 on this duplicate-removed data, including the DESeq2 normalization and subsequent DE computation, in order to obtain a ranked list of differentially abundant OTUs. For RoDEO, projection and scaling is required before the DE computation, in order to make the samples directly comparable across phenotypes. Below is a detailed description of the RoDEO scaling process described in Sect. 2.1.

For the greatest human sample, i.e. the one with smallest number of zeros, we run RoDEO for 100 independent re-sampling simulations, with $P_g = 7$ number of segments, 10^6 number of reads for the re-sampling and gap parameter 1. The number of segments we use to run RoDEO for all the other 96 human samples varies and depends on the result obtained from the scaling process for a given sample. All the other required parameters are instead equal to the ones used for the greatest sample. We then compute the average of projected values for each OTU (average of the 100 iterations), combine all the obtained values in a single matrix and we add to each row i, representing sample i, the difference between the number of segments P_g used to run RoDEO on the greatest sample g and the number of segment P_i used to run RoDEO on sample i.

Similarly, we apply RoDEO projection and the scaling algorithm to the mouse dataset running 100 independent re-sampling simulations, with $P = 10$ number of segments, 10^7 number of reads for the re-sampling and gap 1, for the greatest mouse sample.

Finally, we run RoDEO on the greatest corpse sample for 100 independent re-sampling simulations, with $P = 10$ number of segments, 10^7 number of reads for the re-sampling and gap between the samples equal to 2. In the same way as described before, we compute the averages of projected OTU values for each sample and we add the difference values from the scaling.

References

1. Anastas, P., et al.: 2020 visions. Nature **463**(7277), 26–32 (2010). https://www.nature.com/nature/journal/v463/n7277/full/463026a.html
2. Paulson, J.N., Stine, O.C., Bravo, H.C., Pop, M.: Robust methods for differential abundance analysis in marker gene surveys. Nat. Methods **10**, 1200–1202 (2013)
3. Parida, L., Haiminen, N., Haws, D., Suchodolski, J.: Host trait prediction of metagenomic data for topology-based visualization. In: Natarajan, R., Barua, G., Patra, M.R. (eds.) ICDCIT 2015. LNCS, vol. 8956, pp. 134–149. Springer, Cham (2015). doi:10.1007/978-3-319-14977-6_8
4. Jonsson, V., Österlund, T., Nerman, O., Kristiansson, E.: Statistical evaluation of methods for identification of differentially abundant genes in comparative metagenomics. BMC Genomics **17**(78), 1–14 (2016)
5. Haiminen, N., Klaas, M., Zhou, Z., Utro, F., Cormican, P., Didion, T., Jensen, C., Mason, C.E., Barth, S., Parida, L.: Comparative exomics of Phalaris cultivars under salt stress. BMC Genomics **15**(6), 1–12 (2014)
6. Klaas, M., Haiminen, N., Grant, J., Cormican, P., Finnan, J., Krishna, S., Utro, F., Vellani, T., Parida, L., Barth, S.: Characterizing differentially expressed genes under flooding and drought stress in the biomass grasses Phalaris arundinacea and Dactylis glomerata. Under submission (2017)
7. Karlsson, F.H., Tremaroli, V., Nookaew, I., Bergström, G., Behre, C.J., Fagerberg, B., Nielsen, J., Bäckhed, F.: Gut metagenome in European women with normal, impaired and diabetic glucose control. Nature **498**, 99–103 (2013)
8. Ross, E.M., Moate, P.J., Marett, L.C., Cocks, B.G., Hayes, B.: Metagenomic predictions: from microbiome to complex health and environmental phenotypes in humans and cattle. PLoS ONE **8**, e73056 (2013)
9. Pasolli, E., Tin, D., Truong, F.K., Waldron, L., Segata, N.: Machine learning meta-analysis of large metagenomic datasets: tools and biological insights. PLoS Comput. Biol. **12**(7), e1004977 (2016)
10. Love, M.I., Huber, W., Anders, S.: Moderated estimation of fold change and dispersion for RNA-seq data with DESeq2. Genome Biol. **15**(12), 550 (2014)
11. Weimann, A., Mooren, K., Frank, J., Pope, P.B., Bremges, A., McHardy, A.C., Segata, N.: From genomes to phenotypes: traitar, the microbial trait analyzer. mSystems **1**(6), 1–19 (2016)
12. Ho, T.K.: Random decision forests. In: Proceedings of the Third International Conference on Document Analysis and Recognition, vol. 1, pp. 278–282 (1995)
13. Statnikov, A., Henaff, M., Narendra, V., Konganti, K., Li, Z., Yang, L., Pei, Z., Blaser, M.J., Aliferis, C.F., Alekseyenko, A.V.: A comprehensive evaluation of multicategory classification methods for microbiomic data. Microbiome **1**, 11 (2013)
14. Guyon, I., Elisseeff, A.: An introduction to variable and feature selection. JMLR **3**(11), 57–82 (2013)

15. Metcalf, J.L., Xu, Z.Z., Weiss, S., Lax, S., Van Treuren, W., Hyde, E.R., Song, S.J., Amir, A., Larsen, P., Sangwan, N., Haarmann, D., Humphrey, G.C., Ackermann, G., Thompson, L.R., Lauber, C., Bibat, A., Nicholas, C., Gebert, M.J., Petrosino, J.F., Reed, S.C., Gilbert, J.A., Lynne, A.M., Bucheli, S.R., Carter, D.O., Knight, R.: Microbial community assembly and metabolic function during mammalian corpse decomposition. Science **351**(6269), 158–162 (2016)
16. Caporaso, J.G., Kuczynski, J., Stombaugh, J., Bittinger, K., Bushman, F.D., Costello, E.K., Fierer, N., Gonzalez Peña, A.G., Goodrich, J.K., Gordon, J.I., Huttley, G.A., Kelley, S.T., Knights, D., Koenig, J.E., Ley, R.E., Lozupone, C.A., McDonald, D., Muegge, B.D., Pirrung, M., Reeder, J., Sevinsky, J.R., Turnbaugh, P.J., Walters, W.A., Widmann, J., Yatsunenko, T., Zaneveld, J., Knight, R.: QIIME allows analysis of high-throughput community sequencing data. Nat. Methods **7**(5), 335–336 (2010)

DeepScope: Nonintrusive Whole Slide Saliency Annotation and Prediction from Pathologists at the Microscope

Andrew J. Schaumberg[1,2], S. Joseph Sirintrapun[3],
Hikmat A. Al-Ahmadie[3], Peter J. Schüffler[4],
and Thomas J. Fuchs[3,4]([✉])

[1] Memorial Sloan Kettering Cancer Center and the Tri-Institutional Training
Program in Computational Biology and Medicine, New York, USA
[2] Weill Cornell Graduate School of Medical Sciences, New York, USA
ajs625@cornell.edu
[3] Department of Pathology, Memorial Sloan Kettering Cancer Center,
New York, USA
{sirintrs,alahmadh}@mskcc.org
[4] Department of Medical Physics, Memorial Sloan Kettering Cancer Center,
New York, NY, USA
{schueffp,fuchst}@mskcc.org

Abstract. Modern digital pathology departments have grown to produce whole-slide image data at petabyte scale, an unprecedented treasure chest for medical machine learning tasks. Unfortunately, most digital slides are not annotated at the image level, hindering large-scale application of supervised learning. Manual labeling is prohibitive, requiring pathologists with decades of training and outstanding clinical service responsibilities. This problem is further aggravated by the United States Food and Drug Administration's ruling that primary diagnosis must come from a glass slide rather than a digital image. We present the first end-to-end framework to overcome this problem, gathering annotations in a nonintrusive manner during a pathologist's routine clinical work: (i) microscope-specific 3D-printed commodity camera mounts are used to video record the glass-slide-based clinical diagnosis process; (ii) after routine scanning of the whole slide, the video frames are registered to the digital slide; (iii) motion and observation time are estimated to generate a spatial and temporal saliency map of the whole slide. Demonstrating the utility of these annotations, we train a convolutional neural network that detects diagnosis-relevant salient regions, then report accuracy of 85.15% in bladder and 91.40% in prostate, with 75.00% accuracy when training on prostate but predicting in bladder, despite different pathologists examining the different tissues. When training on one patient but testing on another, AUROC in bladder is 0.79 ± 0.11 and in prostate is 0.96 ± 0.04. Our tool is available at https://bitbucket.org/aschaumberg/deepscope.

© Springer International Publishing AG 2017
A. Bracciali et al. (Eds.): CIBB 2016, LNBI 10477, pp. 42–58, 2017.
DOI: 10.1007/978-3-319-67834-4_4

1   Scientific Background

Computational pathology [12] relies on training data annotated by human experts on digital images. However, the bulk of a pathologist's daily clinical work remains manual on analog light microscopes. A noninterfering system which translates this abundance of expert knowledge at the microscope into labeled digital image data is desired.

Tracking a pathologist's viewing path along the analyzed tissue slide to detect local image saliency has been previously proposed. These approaches include whole slide images displayed on one or more monitors with an eye-tracker [5], mouse-tracker [21] or viewport-tracker [19,23] – but may suffer confounds including peripheral vision [18], head turning [1], distracting extraneous detail [2], monitor resolution [22], multimonitor curvature [25], and monitor bezel field of view fragmentation [27]. Because computer customizations may potentially effect viewing times, for studies of pathologists recorded at a computer, we suggest noting the sensitivity and choice of pointing device, e.g. trackball, touch pad, touch screen, pointing stick, mouse, and if a scroll wheel or keyboard was available to zoom in or out. Only our approach does not change the pathologist's medical practice from the microscope. The microscope is a class I device appropriate for primary diagnosis according to the United States Food and Drug Administration, while whole slide imaging devices are class III [20].

In light of the confounds of alternatives, its centuries of use in pathology, and its favorable regulatory position for primary diagnosis, we believe the microscope is the gold standard for measuring image region saliency. Indeed, there is prior work annotating regions of interest at the microscope for cytology technicians to automatically position the slide for a pathologist [4].

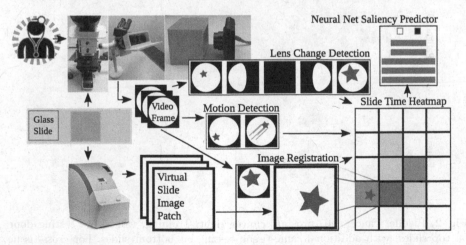

Fig. 1. Proposed microscope-based saliency predictor pipeline workflow. The pathology session is recorded, the slide is scanned, the video frames are registered to scan patches. Lens change detection guides registration and viewing time is recorded for periods without motion. A convolutional neural net learns to classify patches as salient (long looks) or not.

We therefore propose a new, noninterfering workflow for automated video-based detection of region saliency using pathologist viewing time at the microscope (Fig. 1). Viewing time is known in the psychology literature to measure attention [8,15], and we define saliency as pathologist attention when making a diagnosis. Using a commodity digital camera, rather than a custom embedded eye-tracking device [7,16], we video record the pathologist's entire field of view at a tandem microscope to obtain slide region viewing times and register these regions to whole slide image scan regions. Second, we train a convolutional neural network [CNN] on these observation times to predict whether or not a whole slide image region is viewed by a pathologist at the microscope for more than 0.1 seconds [s]. As more videos become available, our CNN predicting image saliency may be further trained and improved, through online learning.

2   Materials and Methods

Pathologists. Pathologists were assistant attending rank with several years experience each. Trainees have different, less efficient, slide viewing strategies [5,18]. Region viewing times and path were automatically recorded during a pathologist's routine slide analysis, without interference.

Patient slides. Two bladder cancer patients were studied by author SJS. Two prostate cancer patients were studied by author HAA. One slide per patient was used, for four slides total (Fig. 2).

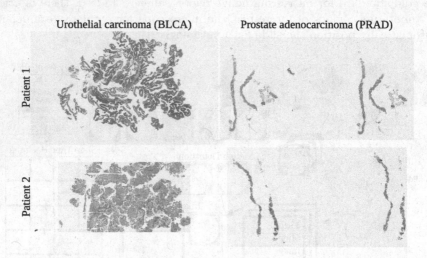

Fig. 2. Bladder cancer left, prostate cancer right. Training, validation, testing done on top slides, with additional same-tissue testing on bottom slides. For cross-tissue testing, top slide tested against other top slide. Viewing time heatmap for top left bladder shown in Fig. 7. Note how the top bladder has more edges than the more solid bottom bladder, while the prostates have similar tissue texture. We believe this impacts interpatient accuracy, shown in Fig. 9.

Fig. 3. Optical flow, showing pixel movement grid. The frame has few moving pixels before *(left)* and after *(right)* pathologist moves the slide. A pathologist looks at a slide region for the duration of consecutive stationary frames.

Scan preprocessing. Microscope slides, inspected by a pathologist, were scanned at 0.5 ± 0.003 microns per pixel [px], using an Aperio AT2 scanner. The resulting SVS data file consists of multiple levels, where level 0 is not downsampled, level 1 is downsampled by a factor of 4, level 2 by a factor of 16, and level 3 by a factor of 32. From each level, 800×800 px patches were extracted via the OpenSlide software library [13]. In bladder, adjacent patches in a level overlap at least 50%, to avoid windowing artifacts in registration. In prostate, adjacent patches overlap at least 75%, to best center the pathologist's field of view on the little tissue in a needle biopsy. Patches evenly cover the entire level without gaps. Scans were either taken before a technician applied marker to the slides, to indicate regions of interest to the pathologist, or after markings were scrubbed from the slide. However, these marks were evident in the pathologist videos discussed in the next section.

Video acquisition. A Panasonic Lumix DMC-FH10 camera with a 16.1 megapixel charge-coupled device [CCD], capable of 720p motion JPEG video at 30 frames per second [fps], was mounted on a second head of an Olympus BX53F multihead teaching microscope to record the pathologist's slide inspection. Microscope objective lens magnifications were 4x, 10x, 20x, 40x, and 100x. Eyepiece lens magnifications was 10x. The pathologist was told to ignore the device and person recording video at the microscope during inspection. The mount (Fig. 1) for this camera was designed in OpenSCAD and 3D-printed on a MakerBot 2 using polylactic acid [PLA] filament.

Camera choice. Many expensive microscope-mounted cameras exist, such as the Lumenera INFINITY-HD and Olympus DP27, which have very good picture quality and frame rate. The Lumenera INFINITY-HD is a CMOS camera, not CCD, so slide movement will skew the image rather than blur it, and we did not want to confound image registration or motion detection with rolling shutter skew. Both cameras trim the field of view to a center-most rectangle for viewing on a computer monitor, which is a loss of information, and we instead assign viewing time to the entire pathologist-viewed 800×800 px PNG patch from the SVS file representing the whole slide scan image. Both cameras do not have USB or Ethernet ports carrying a video feed accessible as a webcam, for registration to the whole slide scan. The Olympus DP27 may be accessible as a Windows TWAIN device, but we could not make this work in Linux. Finally, the HDMI port on both carries high-quality but encrypted video information that we cannot record, and we did not wish to buy a Hauppauge HDMI recording device, because we had a cheaper commodity camera on hand already. We also considered automated screenshots of the video feed in Aperio ImageScope as displayed on a computer monitor, but we observed a lower frame rate and detecting lens change is complicated because the entire field of view is not available. Recording low-quality video on a commodity camera to a SecureDigital [SD] memory card is inexpensive, captures the entire field of view, and is generally applicable in any hospital. For this pilot study, we used only one camera for video recording, rather than two different microscope cameras, potentially eliminating a confound for how many pixels are moving during rapid short movements of the slide. For 3D printing requisite camera mounts, open source tools are available.

Video preprocessing and registration. A Debian Linux computer converted individual slide inspection video frames to PNG files using the ffmpeg program. OpenCV software detected slide movement via a dense optical flow procedure [9,10], comparing the current and preceding video frames, shown in Fig. 3. Through this dense optical flow procedure. We calculated a movement vector for each pixel of each camera video frame, where a movement vector magnitude of one means the pixel has been displaced by one pixel in the video frame of interest,

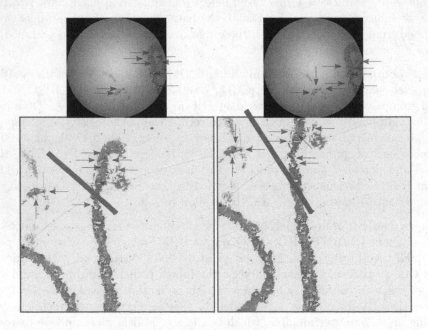

Fig. 4. The best image registration for a given video frame (same frame top left and top right) from the commodity camera at the microscope eyepiece compared to two different high-quality patches (bottom left and bottom right) from the whole slide scan image minimizes the length of the green line, which is the distance from the center of the patch to the center of the frame mapped into the patch's coordinate space. The green line's length is distance d in Algorithm 1. (Color figure online)

> **input** : I_{frame}: image from commodity camera, a video frame
> $I_{0,1,...,N-1}$: N overlapping patch images, together spanning whole slide
> $I_{prior} \in I_{0,1,...,N-1}$, the best matching patch from previous video frame
> **output**: $I_{best} \in I_{0,1,...,N-1}$, the best matching patch to I_{frame}
> $S_{frame} \longleftarrow$ set of all SURF interest points in I_{frame};
> $n \longleftarrow 0$, a counter through $I_{0,1,...,N-1}$ images;
> $n_{best} \longleftarrow -1$, the value of n where I_n is I_{best};
> $d_{best} \longleftarrow MAXINT$, to store the distance between I_{best} and I_{frame} centers;
> **while** $n < N$ **do**
> > **if** I_n *is three or fewer patches spatially removed from* I_{prior} **then**
> > > $S_t \longleftarrow$ set of all SURF interest points in $I_n \in I_{0,1,...,N-1}$;
> > > $S_{fs} \longleftarrow$ subset of S_{frame} points that match SURF feature vector of an S_t point;
> > > $S_{ts} \longleftarrow$ subset of S_t points that match SURF feature vector of an S_{frame} point;
> > > $T \longleftarrow$ rigid body transformation of I_{frame} pixel coordinate space into I_n pixel coordinate space, calculated by point set registration of $RANSAC(S_{fs}, S_{ts})$;
> > > $d \longleftarrow$ distance in pixels between I_{frame} center and $T(I_{frame})$ center, which measures how far off-center I_n is from I_{frame};
> > > **if** $d < d_{best}$ **then**
> > > > $n_{best} \longleftarrow n$;
> > > > $d_{best} \longleftarrow d$;
> > > **end**
> > **end**
> > $n \longleftarrow n + 1$;
> **end**
> return $I_{n \longleftarrow n_{best}}$, which is I_{best};

Algorithm 1. Automated image registration procedure (Fig. 4) to find the least off-center patch from a given commodity camera video frame. The whole slide image is split into N overlapping 800×800 px patches. "Three or fewer patches spatially removed" means any I_n must be (i) I_{prior}, (ii) adjacent to I_{prior}, (iii) adjacent to a patch adjacent to I_{prior}, or (iv) adjacent to a patch adjacent to an I_{prior}-adjacent patch. In this way, I_n is restricted to a spatial neighborhood localized around the prior match, typically improving image registration performance because most slide movements are small. On lens change, (i) the patch at lower magnification and (ii) the patches at higher magnification covering the same area as the current magnification's neighborhood are considered for registration only.

with respect to the previous video frame. Though the details of this procedure are beyond the scope of this work, a computationally efficient polynomial expansion method explains a pixel's movement vector as the previous frame's pixel neighborhood polynomial transformed under translation to the current frame's pixel neighborhood polynomial, where a 39×39 px Gaussian weighting function averages pixel movement vectors for smoothing [9,10]. We defined slide movement to start if 10% or more of pixels in the entire field of view of the camera have a movement vector magnitude of at least one, and defined slide movement to stop if 2% or fewer of the pixels in the entire field of view of the camera have a movement magnitude vector of at least one. The entire field of view of the camera is 640×480 px, and a small subset of these capture the circular field of view at the microscope eyepiece, with the remaining pixels being black (Fig. 3). The representative frame among consecutive unmoving frames moved the least. The ImageJ [24] SURF [3][1] and OpenCV software libraries registered each representative to an 800×800 px image patch taken from the high-resolution Aperio slide scanner. Each patch aggregated total pathologist time.

The partially automated registration process starts with initial manual registration of a frame, followed by automated registration within the preceding registration's spatial neighborhood (Fig. 4 and Algorithm 1). First, (i) a set S_{frame} of SURF interest points were found in the video frame, (ii) a set S_t of SURF interest points were found in a slide image patch, (iii) SURF interest point feature vectors were compared in S_{frame} and S_t to determine which points were shared in S_{frame} and S_t, and (iv) subsets of S_{frame} and S_t points that were shared were then stored in S_{fs} and S_{ts}, respectively. Points shared between a camera video frame and an image patch (Fig. 4 at left, top and bottom) change depending on the image patch (Fig. 4 at right, top and bottom). Second, we used the OpenCV implementation of random sample consensus [RANSAC] [11] for point set registration, to calculate a rigid body transformation from S_{fs} point pixel positions in the video frame to S_{ts} point pixel positions in the image patch, to find the distance in pixels that the video frame is off-center from the patch. Following this procedure for every image patch in the spatial neighborhood of the previous image registration, we selected the least off-center image patch as the best registration, because the pathologist's fovea is in approximately the same place in this video frame and image patch. Finally, a manual curation of registrations ensures correctness. Because slide movements are usually slight, this automated process reduces manual curation effort because automatic registrations are rarely far from the correct registration, so after the registration is corrected within a small localized neighborhood, automatic registrations may proceed from there. Fully automated image registration is not part of this study.

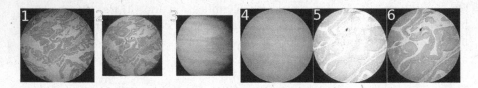

Fig. 5. Lens change detection: the normal non-black pixel bounding box is initially 415×415 px. A change to 415×282 px indicates the pathologist changing the lens, thus changing slide magnification. Note some pixels that may appear black are called non-black due to difficult to perceive noise in the image, which effects calculated bounding box size. All images shown at same scale trimmed to bounding box.

---

[1] ImageJ SURF is released under the GNU GPL and is available for download from http://labun.com/imagej-surf/.

Slide magnification may change during inspection as the pathologist changes objective lenses. Lens change is detected automatically when the field of view bounding box of nonblack pixels changes size (Fig. 5). SURF is scale-invariant so registrations may otherwise proceed at an unchanged magnification.

Deep learning. We used Caffe [14] for deep learning of convolutional features in a binary classification model given the 800×800 px image patches labeled with pathologist viewing times in seconds. To adapt for our purpose CaffeNet (Fig. 6), which is similar to AlexNet [17], we re-initialized its top layer's weights after ImageNet [6] pre-training. Two output neurons were connected to the re-initialized layer, then training followed on augmented 800×800 px patches for 10,000 iterations in Caffe. In bladder, our model simply predicted whether or not a pathologist viewed an 800×800 px patch more than 0.1 s (30 fps camera). In prostate, due to the higher overlap between adjacent patches and less tissue available, to be salient a patch met at least one of these criteria: (1) viewed more than 0.1 s, (2) immediately above, below, left, or right of at least two patches viewed more than 0.1 s, or (3) above, below, left, right, or diagonal from at least three patches viewed more than 0.1 s such that all three are not on the same side. In this way, image patches highly overlapping in the neighborhood (Algorithm 1) of salient patches were not themselves considered nonsalient if a pathologist happened to jump over them during observation.

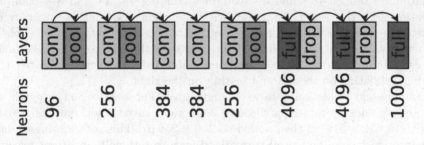

Fig. 6. Caffenet neuron counts, convolutional layers, pooling layers, dropout [26] layers, and fully-connected layers.

3   Experiments

Urothelial carcinoma (bladder) in Fig. 7 was analyzed first, with author HAA inspecting at the microscope. Viewed regions at the microscope corresponded to the whole-slide scan SVS file at magnification levels 2 and 1. We restricted our analysis to level 2, having insufficient level 1 data. We split level 2 into three portions: left, center, and right. Due to over 50% overlap among the slide's total 54 800 $\times$ 800 px level 2 patches, we excluded the center portion from analysis, but retained left and right sides, which did not overlap (Fig. 8).

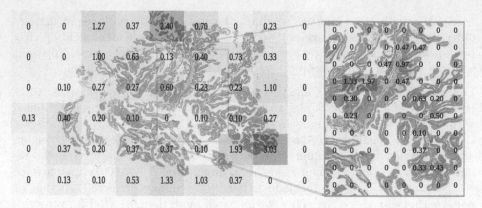

Fig. 7. Pathologist viewing times in seconds at the microscope for low (*left*, 10x, level 2) and high magnification (*right*, 20x, level 1), registered to the same urothelial carcinoma slide scan.

In bladder, we considered a negative example to be a patch viewed for 0.1 s (3 frames or fewer, 30 fps) or less, and a positive example viewed for more than 0.1 s (4 frames or more). This produced 9 positive and 9 negative examples on the left side, and the same number on the right. We performed three-fold cross-validation on the left (6+ and 6− examples training set, 3+ and 3− examples validation set), then used the model with the highest validation accuracy on the right to calculate test accuracy, an estimate of generalization error (Fig. 9). This cross-validation was duplicated ten times on the left, each time estimating test accuracy, to calculate a confidence interval. We then duplicated this training/validating on the left and testing on the right.

Training and validation data were augmented. For a 800 × 800 px patch, all flips and one-degree rotations through 360° were saved, then cropped to the centermost 512 × 512 px, then scaled to 256 × 256 px. This rotation-based data augmentation biases the neural network to learn rotationally-invariant features rather than overfit to the training data's particular orientation, e.g. the angle of prostate needle biopsy tissue strips. Thus intrapatient and interpatient test sets are not augmented, but training and validation sets are augmented. The cancer diagnosis or viewing time in pathology is not expected to change when rotating or flipping a slide. We direct readers to Krizhevsky *et al.* 2012 [17] for more information on data augmentation. Like Krizhevsky's data augmentation of 224 × 224 px random crops for small translations, we further augment our dataset through random crops of 227 × 227 px, which is the default for CaffeNet. Unlike Krizhevsky, we do not augment our dataset through minor perturbations in the principal components of the RGB color space.

In bladder, the augmented training set size was 8,640 patches. This 8,640 count includes rotations and flips, but does not include random crops, which were performed automatically by Caffe at training time. Caffe randomly cropped 256 × 256 px patches to 227 × 227 px for each iteration of CaffeNet learning. No images in the validation set were derived from the training set, and vice versa. A training set consists of two concatenated folds, with the remaining fold as validation. In addition to the bladder cancer slide, we analogously processed two prostate cancer needle biopsy slides, with author SJS inspecting these slides. In prostate, the augmented training set size was 8,160 patches.

Training and validation sets are drawn from the same side of the slide, i.e. both sets on the left or both sets on the right (Fig. 8). Patches in a training set may have at least 50% overlap with patches in a validation set. Overlapping regions of these images have identical sets of pixels, guaranteeing training and validation sets are exchangeable for valid cross-validation. If training error steadily decreases while validation error steadily increases, where training and validation sets are exchangeable, then the classifier is overfit. In contrast, the other side of the slide is used as a test set and may appear obviously different

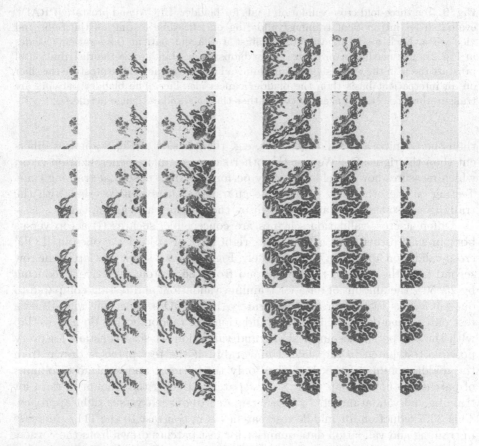

Fig. 8. Scaled image patches of left and right sides of bladder patient 1 slide (Fig. 2). Middle excluded here and not used in analysis, to isolate left and right sides from each other. Note far left and far right have less tissue, but tissue is present for training. The overlap among patches is evenly distributed and greater than 50%.

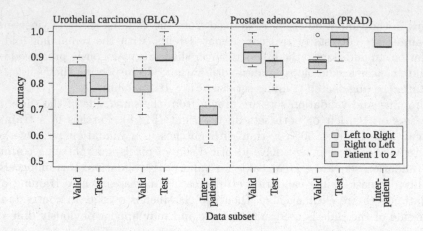

Fig. 9. Ten three-fold cross-validation trials for bladder [BLCA] and prostate [PRAD], evaluated for intrapatient training/validating on left while testing on the right and vice versa. Each model is evaluated against a different patient (interpatient), slides in Fig. 2). The needle for prostate cancer biopsy may standardize the distribution of prostate tissue in the whole slide, maintaining a higher accuracy of the prostate classifier on an interpatient basis than the bladder cancer classifier. The bladder patients are transurethral resections taken by cuts rather than a standard gauge needle.

than the training and validation sets, e.g. the left side of Fig. 8 appears different than the right side. We test the other side to estimate generalization error, which measures how the classifier may perform on data unseen at training time. Testing on the other side of the slide guarantees there is no overlap with the training set, so the test data is unseen by the classifier at training time.

Different cross-validation schemes are conceivable, such as (i) a top versus bottom split rather than a left versus right split or (ii) a leave-one-out [LOO] cross-validation approach. Unfortunately, Fig. 8 shows a slight overlap in the row second from the top and the row second from the bottom, effectively reducing by 25–50% the amount of data for training, validation, and testing compared to our left versus right approach. Separately, in a LOO setting, one may draw a test patch, then draw training and validation sets randomly that do not overlap with the test patch, keeping training and validation set sizes constant for every possible test patch in the slide. Unfavorably, if the test patch is drawn from the middle column of the slide, then only the leftmost and rightmost columns of patches do not overlap with the test patch, reducing the amount of data for training and validation sets by 33% compared to our left versus right approach. This 33% reduction for middle test patches is in contrast to the 111% increase in training and validation data quantity for test patches drawn from the corners of the leftmost or rightmost columns, where this excess is randomly discarded to maintain constant training and validation set sizes for all possible test patches. Moreover, if the test patch is in the bottom row on the right side, the top row on the right side may be sampled for training, which may inflate the LOO generalization accuracy estimate compared to our cross-validation approach that trains only on the left when testing on the right, due to patches on the right appearing similar to one another. We show in Sect. 4 that training on the left and testing on the right gives significantly different accuracy compared to training on the right and testing on the left, suggesting the left and right sides have indeed different distributions of information. Thus compared to these alternatives, our left versus right three-fold cross-validation approach (i) maximizes the sizes of the training, validation, and test sets, (ii) conservatively estimates generalization error by not training the classifier on data that appear similar to the test set, and (iii) samples each patch on the left or right sides exactly once for an overall validation error measure for that side.

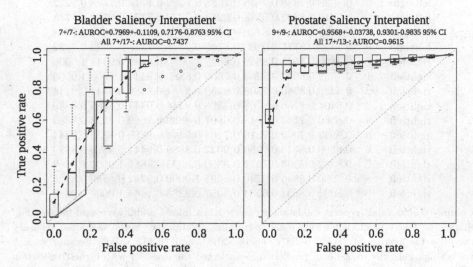

Fig. 10. Interpatient area under the receiver operating characteristic [AUROC] for bladder and prostate, with dashed black curve for average AUROC over draws of the data and blue line for all data used from the patient. (Color figure online)

4 Results

In bladder, when training/validating on the left side and testing on the right, mean test accuracy is 0.781 ± 0.0423 (stdev) with 95% confidence interval [CI] from 0.750 to 0.811 ($df = 9$, Student's T, Table 1). When training/validating on the right and testing on the left, mean test accuracy is 0.922 ± 0.0468 with $0.889 - 0.956$ 95% CI (Table 1). Overall mean test accuracy is 85.15%. The left and right test accuracies differ ($p = 0.000135$, Wilcoxon rank-sum, $n = 20$), while validation accuracies do not ($p = 0.9118$, $n = 20$). This suggests nonhomogenous information content throughout the slide. Indeed, the pathologist started and ended slide inspection on the right, and spent double the time on the right versus the left (Fig. 7, 8.32 s right, 4.07 s left). The second bladder had different

Table 1. Accuracies of ten trials of three-fold cross-validation in bladder. Validation and test accuracies for a single slide video of urothelial carcinoma (patient 1, slide at upper left in Fig. 2, performance plotted at left in Fig. 9), left side of the slide versus right side.

Direction	Trial	Fold0 Valid	Fold1 Valid	Fold2 Valid	Valid Acc	Fold0 Test	Fold1 Test	Fold2 Test	Test Acc
leftright	0	0.9466	0.5850	0.9680	0.8332	0.8333	0.7222	0.7778	0.7778
leftright	1	0.8852	0.9070	0.9070	0.8997	0.7778	0.7222	0.7778	0.7500
leftright	2	0.9218	0.8602	0.8832	0.8884	0.8333	0.7222	0.7778	0.8333
leftright	3	0.7640	0.7812	0.7120	0.7524	0.7778	0.7778	0.7778	0.7778
leftright	4	0.7590	0.6576	0.5134	0.6433	0.8333	0.7778	0.7778	0.8333
leftright	5	0.9268	0.7416	0.5088	0.7257	0.8333	0.7778	0.7778	0.8333
leftright	6	0.8028	0.7988	0.7048	0.7688	0.7222	0.7778	0.7778	0.7222
leftright	7	0.7318	0.8402	0.9088	0.8269	0.7778	0.7778	0.7778	0.7778
leftright	8	0.9572	0.7608	0.8418	0.8533	0.7778	0.8333	0.8333	0.7778
leftright	9	0.7492	0.8774	0.9860	0.8709	0.7778	0.7778	0.7222	0.7222
rightleft	0	0.8802	0.8528	0.8554	0.8628	1.0000	0.8889	0.9444	1.0000
rightleft	1	0.7662	0.5982	0.9364	0.7669	0.8889	0.9444	1.0000	1.0000
rightleft	2	0.9492	0.8560	0.7308	0.8453	0.9444	0.8889	0.9444	0.9444
rightleft	3	0.5404	0.8206	0.8368	0.7326	0.9444	0.9444	0.8889	0.8889
rightleft	4	0.6560	0.7114	0.6748	0.6807	0.8889	0.8889	0.8333	0.8889
rightleft	5	0.8932	0.7062	0.7310	0.7768	0.9444	0.8889	0.8333	0.9444
rightleft	6	0.8560	0.8540	0.9966	0.9022	0.8889	0.9444	0.8889	0.8889
rightleft	7	0.8362	0.8560	0.7978	0.8300	0.8333	0.8889	1.0000	0.8889
rightleft	8	0.7200	0.8546	0.9740	0.8495	1.0000	0.9444	0.8889	0.8889
rightleft	9	0.8634	0.8634	0.6904	0.8057	0.8333	0.9444	1.0000	0.8889

Column "Fold0 Valid" reports validation accuracy when folds 1 and 2 were used for training. Similarly, "Fold1 Valid" is for folds 0 and 2 training. "Valid Acc" is the validation accuracy overall – the average of "Fold0 Valid", "Fold1 Valid", and "Fold2 Valid". Because we will use a single classifier for saliency prediction, we selected the classifier with highest validation accuracy and highlighted it yellow, e.g. we selected the Fold0 classifier with 0.9218 validation error as shown in the third row, namely Trial 2 leftright.

Column "Fold0 Test" reports the test accuracy of the classifier trained on folds 1 and 2. Because we will use a single classifier not an ensemble, we highlight the test accuracy of the classifier selected by highest validation accuracy and report this in "Test Acc" as generalization accuracy, e.g. we selected the Trial 2 leftright Fold0 classifier having "Fold0 Test" of 0.8333 and copied this to "Test Acc". We report test accuracies for all three classifiers, showing Fold1 and Fold2 classifiers tie for highest validation accuracy in Trial 1 leftright, so their test accuracies of 0.7222 and 0.7778 were averaged for a "Test Acc" of 0.7500. As another sanity check in our small data setting, we report that the variance in the selected-versus-non-selected test accuracy differences is not greater than the selected-versus-non-selected validation accuracy differences (F-Test $p = 0.5662$ and Bartlett's Test $p = 0.5661$. Validation differences normally distributed by Anderson-Darling $p = 0.08837$, and test differences by $p = 0.1734$). If it were greater, there may be experimental setup problems because training would not be stably producing classifiers that learn the saliency concept. Finally, one may train a classifier on all folds then evaluate test accuracy with this classifier, but a performance boost from additional training data may inflate generalization accuracy. In Sect. 4 we show without such inflation there remains a significant difference in generalization accuracy and interpatient accuracy in bladder.

Testing the best classifier (highlighted in cyan, highest test accuracy on this and other folds, secondarily highest mean validation accuracy) on draws of the data on the second bladder patient, accuracies are 0.643, 0.786, 0.714, 0.786, 0.714, 0.714, 0.643, 0.571, 0.643, and 0.571.

Table 2. Accuracies of ten trials of three-fold cross-validation in prostate. Validation and test accuracies for a single slide video of prostate adenocarcinoma (patient 1, slide at upper right in Fig. 2, performance plotted at right in Fig. 9), left side of the slide versus right side.

Direction	Trial	Fold0 Valid	Fold1 Valid	Fold2 Valid	Valid Acc	Fold0 Test	Fold1 Test	Fold2 Test	Test Acc
leftright	0	0.9992	0.9946	1.0000	0.9979	0.7778	0.7778	0.7778	0.7778
leftright	1	0.9512	0.7282	0.9994	0.8929	0.8889	0.8889	0.8889	0.8889
leftright	2	0.9550	1.0000	0.6530	0.8693	0.8889	0.7778	0.7222	0.7778
leftright	3	0.8636	0.9992	1.0000	0.9543	0.8889	0.7778	0.9444	0.9444
leftright	4	0.8276	0.7760	0.9940	0.8659	0.8889	0.7778	0.8889	0.8889
leftright	5	0.8654	0.9986	1.0000	0.9547	0.9444	0.9444	0.9444	0.9444
leftright	6	0.8560	0.9862	0.9992	0.9471	0.8889	0.8889	0.8889	0.8889
leftright	7	0.8674	1.0000	0.9984	0.9553	0.8889	0.8333	0.8889	0.8333
leftright	8	0.9560	0.8560	0.8760	0.8960	0.8889	0.8333	0.9444	0.8889
leftright	9	0.6846	0.9992	0.9560	0.8799	0.8333	0.8333	1.0000	0.8333
rightleft	0	0.9786	0.7760	0.9146	0.8897	0.8889	0.9444	1.0000	0.8889
rightleft	1	1.0000	0.7292	0.8460	0.8584	1.0000	0.9444	1.0000	1.0000
rightleft	2	0.7130	0.9512	0.8676	0.8439	0.8889	1.0000	0.8333	1.0000
rightleft	3	0.9998	1.0000	0.9664	0.9887	0.9444	0.9444	1.0000	0.9444
rightleft	4	0.7760	1.0000	0.8842	0.8867	0.8889	0.9444	1.0000	0.9444
rightleft	5	0.9758	0.9984	0.5926	0.8556	0.9444	0.8889	0.9444	0.8889
rightleft	6	0.6344	0.9770	1.0000	0.8705	0.8889	1.0000	1.0000	1.0000
rightleft	7	0.7760	0.9028	1.0000	0.8929	0.8889	1.0000	1.0000	1.0000
rightleft	8	0.8560	0.8560	0.9992	0.9037	0.9444	0.9444	0.9444	0.9444
rightleft	9	0.8560	0.9412	0.8538	0.8837	1.0000	1.0000	1.0000	1.0000

Testing the best classifier on draws of the data on the second prostate patient, accuracies are 0.944, 1, 0.944, 0.944, 1, 0.944, 0.944, 0.944, 1, and 1.

morphology and model accuracy reduced to 0.678 ± 0.0772, $0.623 - 0.73495\%$ CI. Moreover, the second bladder had only 7 positive examples available, whereas both prostates and the first bladder had at least 9 positive examples.

For the first prostate slide, training on the left side and testing on the right, we find accuracy 0.867 ± 0.0597, $0.824 - 0.909$ 95% CI (Table 2). Training on the right and testing on left, we find 0.961 ± 0.0457, $0.928 - 0.994$ 95% CI (Table 2). Overall mean test accuracy is 91.40%. Taking the best model learned from this first prostate (right side, test accuracy 100%, 18/18), we tested on the second prostate's right side (because the left did not have 9 positive training examples) and find 0.967 ± 0.0287, $0.946 - 0.987$ 95% CI. We also tested this model on the bladder cancer slide, and find 0.780 accuracy on the left and 0.720 on the right (9+ and 9− training examples each), mean accuracy 75.00%. The best bladder cancer model predicts every patch is not salient in both prostates, presumably because the little tissue in prostate is insufficient for a positive saliency prediction.

Interpatient AUROC for bladder and prostate is shown in Fig. 10. In prostate, nine salient and nine nonsalient examples are drawn from the second patient.

Average AUROC was calculated from ten such draws, achieving a mean ± stdev of 0.9568 ± 0.0374 and 95% CI of 0.9301 − 0.9835. Over all 17 salient and 13 nonsalient patches used from the second prostate patient, the AUROC is 0.9615. In bladder, due to fewer patches available in the small slide, only seven salient and seven nonsalient examples are drawn from the second patient. Average AUROC was calculated for ten such draws, achieving 0.7929 ± 0.1109 and 95% CI of 0.7176 − 0.8763. Over all 7 salient and 17 nonsalient patches used from the second bladder patient, the AUROC is 0.7437. These nonoverlapping confidence intervals are evidence the bladder cancer classifier distinguishes salient from nonsalient patches less well than the prostate cancer classifier, and a Wilcoxon rank-sum test indeed finds the difference in classifier performance by these ten draws each from bladder and prostate is significant ($p = 0.0001325$) (Fig. 9).

The deep convolutional network CaffeNet emits a score from 0 to 1 when predicting if an image patch is salient or not. When taking a score of greater than 0.5 to be salient, the p-value from Fisher's Exact Test is 1.167e–7 in prostate (16 true positives, 1 false negative, 0 false positives, 13 true negatives) and 0.009916 in bladder (7 true positives, 0 false negatives, 7 false positives, 10 true negatives), indicating our trained CaffeNet classifier in both tissues accurately distinguishes salient from nonsalient regions when trained on one patient and predicting in another.

5   Conclusion

Collecting image-based expert annotations for the deluge of medical data at modern hospitals is one of the tightest bottlenecks for the application of large-scale supervised machine learning. We address this with a novel framework that combines a commodity camera, 3D-printed mount, and software stack to build a predictive model for saliency on whole slides, i.e. where a pathologist looks to make a diagnosis. The registered regions from the digital slide scan are markedly higher quality than the camera frames, since they do not suffer from debris, vignetting, and other artifacts. The proposed CNN is able to predict salient slide regions with a test accuracy of 85–91%. We plan to scale up this pilot study to more patients, tissues, and pathologists.

Acknowledgments. AJS was supported by NIH/NCI grant F31CA214029 and the Tri-Institutional Training Program in Computational Biology and Medicine (via NIH training grant T32GM083937). This research was funded in part through the NIH/NCI Cancer Center Support Grant P30CA008748. AJS thanks Terrie Wheeler, Du Cheng, and the Medical Student Executive Committee of Weill Cornell Medical College for free 3D printing access, instruction, and support. AJS thanks Mariam Aly for taking the photo of the camera on the orange 3D-printed mount in Fig. 1, and attention discussion. We acknowledge fair use of part of a doctor stick figure image in Fig. 1 from 123rf.com. AJS thanks Mark Rubin for helpful pathology discussion. AJS thanks Paul Tatarsky and Juan Perin for Caffe install help on the Memorial Sloan Kettering supercomputer. We gratefully acknowledge NVIDIA Corporation for providing us a GPU as part of the GPU Research Center award to TJF, and for their support with other GPUs.

References

1. Ball, R., North, C.: The effects of peripheral vision and physical navigation on large scale visualization. In: Proceedings of Graphics Interface, pp. 9–16 (2008)
2. Ball, R., North, C., Bowman, D.: Move to improve: promoting physical navigation to increase user performance with large displays, pp. 191–200. ACM (2007)
3. Bay, H., Tuytelaars, T., Van Gool, L.: SURF: speeded up robust features. In: Leonardis, A., Bischof, H., Pinz, A. (eds.) ECCV 2006. LNCS, vol. 3951, pp. 404–417. Springer, Heidelberg (2006). doi:10.1007/11744023_32
4. Begelman, G., Lifshits, M., Rivlin, E.: Visual positioning of previously defined ROIs on microscopic slides. IEEE Trans. Inf. Technol. Biomed. $10(1)$, 42–50 (2006)
5. Brunye, T., Carney, P., Allison, K., Shapiro, L., Weaver, D., Elmore, J.: Eye movements as an index of pathologist visual expertise: a pilot study. PLoS ONE $9(8)$, e103447 (2014)
6. Deng, J., Dong, W., Socher, R., Li, L.J., Li, K., Fei-Fei, L.: ImageNet: a large-scale hierarchical image database, pp. 248–255. IEEE, June 2009
7. Eivazi, S., Bednarik, R., Leinonen, V., von und zu Fraunberg, M., Jaaskelainen, J.: Embedding an eye tracker into a surgical microscope: requirements, design, and implementation. IEEE Sens. J. $16(7)$, 2070–2078 (2016)
8. Erwin, D.: The Interface of Language, Vision, and Action. Routledge, London (2004). doi:10.4324/9780203488430
9. Farneback, G.: Polynomial expansion for orientation and motion estimation. Ph.D. thesis, Linkoping University, Sweden (2002)
10. Farnebäck, G.: Two-frame motion estimation based on polynomial expansion. In: Bigun, J., Gustavsson, T. (eds.) SCIA 2003. LNCS, vol. 2749, pp. 363–370. Springer, Heidelberg (2003). doi:10.1007/3-540-45103-X_50
11. Fischler, M., Bolles, R.: Random sample consensus: a paradigm for model fitting with applications to image analysis and automated cartography. Commun. ACM $24(6)$, 381–395 (1981)
12. Fuchs, T., Buhmann, J.: Computational pathology: challenges and promises for tissue analysis. Comput. Med. Imaging Graph. $35(7–8)$, 515–530 (2011). The official journal of the Computerized Medical Imaging Society
13. Goode, A., Gilbert, B., Harkes, J., Jukic, D., Satyanarayanan, M.: OpenSlide: a vendor-neutral software foundation for digital pathology. J. Pathol. Inform. 4, 27 (2013)
14. Jia, Y., Shelhamer, E., Donahue, J., Karayev, S., Long, J., Girshick, R., Guadarrama, S., Darrell, T.: Caffe: convolutional architecture for fast feature embedding, June 2014
15. Just, M., Carpenter, P.: A theory of reading: from eye fixations to comprehension. Psychol. Rev. $87(4)$, 329–354 (1980)
16. Keerativittayanun, S., Rakjaeng, K., Kondo, T., Kongprawechnon, W., Tungpimolrut, K., Leelasawassuk, T.: Eye tracking system for ophthalmic operating microscope, pp. 653–656. IEEE, August 2009
17. Krizhevsky, A., Sutskever, I., Hinton, G.: Imagenet classification with deep convolutional neural networks (2012)
18. Krupinski, E., Tillack, A., Richter, L., Henderson, J., Bhattacharyya, A., Scott, K., Graham, A., Descour, M., Davis, J., Weinstein, R.: Eye-movement study and human performance using telepathology virtual slides. Implications for medical education and differences with experience. Hum. Pathol. $37(12)$, 1543–1556 (2006)

58   A.J. Schaumberg et al.

19. Mercan, E., Aksoy, S., Shapiro, L., Weaver, D., Brunye, T., Elmore, J.: Localization of diagnostically relevant regions of interest in whole slide images, pp. 1179–1184. IEEE, August 2014
20. Parwani, A., Hassell, L., Glassy, E., Pantanowitz, L.: Regulatory barriers surrounding the use of whole slide imaging in the United States of America. J. Pathol. Inform. **5**(1) (2014)
21. Raghunath, V., Braxton, M., Gagnon, S., Brunye, T., Allison, K., Reisch, L., Weaver, D., Elmore, J., Shapiro, L.: Mouse cursor movement and eye tracking data as an indicator of pathologists' attention when viewing digital whole slide images. J. Pathol. Inform. **3**, 43 (2012)
22. Randell, R., Ambepitiya, T., Mello-Thoms, C., Ruddle, R., Brettle, D., Thomas, R., Treanor, D.: Effect of display resolution on time to diagnosis with virtual pathology slides in a systematic search task. J. Digit. Imaging **28**(1), 68–76 (2015)
23. Romo, D., Romero, E., Gonzalez, F.: Learning regions of interest from low level maps in virtual microscopy. Diagn. Pathol. **6**(Suppl 1), S22 (2011)
24. Schneider, C., Rasband, W., Eliceiri, K.: NIH image to ImageJ: 25 years of image analysis. Nat. methods **9**(7), 671–675 (2012)
25. Shupp, L., Ball, R., Yost, B., Booker, J., North, C.: Evaluation of viewport size and curvature of large, high-resolution displays, pp. 123–130. Canadian Information Processing Society (2006)
26. Srivastava, N., Hinton, G., Krizhevsky, A., Sutskever, I., Salakhutdinov, R.: Dropout: a simple way to prevent neural networks from overfitting, vol. 15, pp. 1929–1958, June 2014
27. Starkweather, G.: 58.4: DSHARP–a wide screen multi-projector display. SID Symp. Digest Tech. Pap. **34**(1), 1535–1537 (2003)

PLS-SEM Mediation Analysis
of Gene-Expression Data for the Evaluation
# of a Drug Effect

Daniele Pepe[(⊠)] and Tomasz Burzykowski

I-BioStat, Hasselt University, Campus Diepenbeek, Hasselt, Belgium
daniele.pepe@uhasselt.be

Abstract. Gene expression analysis can unveil the genes associated with the molecular action of a drug. However, it is not always clear how the differentially expressed genes restore the phenotype and whether, globally, the drug has an effect on the disease. We propose a method that exploits gene-expression data and network biology information to build a mediation analysis model for the evaluation of the effect of treatment on the disease at molecular level. First, differentially expressed genes (DEGs) associated to the drug and the disease are discovered. Then, based on a pathway analysis, shortest paths between drug DEGs and disease DEGs are obtained. This allows discovering the mediator genes that connect drug genes to disease genes. The expression values of the three sets of genes are used to conduct a mediation analysis that evaluates the effect of the drug on the disease. The effect could be direct, indirect by mediators, or both. The latent variables and mediation model are constructed by using the PLS-SEM. The procedure is applied to a real example concerning the effect of abacavir on HIV samples. The proposed pipeline can offer an additional tool for the understanding of the etiology of a disease and unveiling the mechanisms of action of a drug at gene level.

Keywords: PLS-SEM · Mediation model · Gene expression · Network analysis

1 Introduction

The use of gene expression data can be extremely valuable in the understanding of the molecular mechanisms related to a disease and its treatment. The gene-expression profiling could provide clues about regulatory mechanisms, biochemical pathways, and cellular function. In addition, the determination of genes expressed in untreated disease samples, as compared to samples treated with a particular drug, could provide a way to understand the mechanism of the action of a drug [1]. Often, the analysis based on expression profiling consists of comparing the separate gene profiles of the disease and of the drug, with a subsequent biological interpretation aimed at connecting the two lists. The analysis could be conducted at the gene level or at the pathway level. However, this approach suffers from two important limitations: (1) it ignores the way in which the lists of genes are connected from a network point of view [2]; (2) it does not

© Springer International Publishing AG 2017
A. Bracciali et al. (Eds.): CIBB 2016, LNBI 10477, pp. 59–69, 2017.
DOI: 10.1007/978-3-319-67834-4_5

offer a method to confirm whether the treatment has got a statistically significant effect on the disease.

Mediation analysis found numerous applications in basic and applied research. The goal of the analysis is to discover how an independent variable X acts on a dependent variable Y. There are four possible scenarios: (1) X has not effect on Y; (2) it is easily understandable that mediation analysis is perfectly suitable for the evaluation of pharmacological and psychotherapeutic treatments. To perform a mediation analysis, a causal inference analysis is necessary. The classical approach, proposed by Baron and Kenny [3], is based on a series of regression analyses to evaluate the significance of the direct and the indirect effects. However, the approach does not consider the intrinsic causal nature of the model. In fact, in the multiple regression analysis context, a causal model is constructed by a series of regression equations with each regressor estimated and evaluated separately. Structural equation modeling (SEM) overcomes this problem by fitting the causal model entirely in one step and evaluating it by appropriate goodness-of-fit indices [4]. This not only permits to estimate the direct and the indirect effects, but also to verify if the mediation model fits the observed data. Furthermore, mediation analysis in the SEM framework can include latent variables and the evaluation of more complicated mediation models with, for example, more than one mediator [5].

In this paper, we consider an approach that uses differentially expressed genes (DEGs) associated with the disease and with the effect of the treatment on the disease to: (1) investigate how, from a molecular point of view, the genes are connected; (2) infer whether the treatment has a statistically significant effect on the disease from a gene expression point of view. The first goal is addressed by searching for the directed shortest paths between the drug DEGs and the disease DEGs based on the KEGG pathways [6, 7]. This allows building a network where the source nodes are the drug DEGs, the destination nodes are the disease DEGs, and the genes that connect both sets of nodes are the mediator genes. The second goal is reached by using mediation analysis with PLS-SEM [8], a statistical approach particularly suitable for the creation of latent variables and the analysis of their relationships. PLS-SEM is different from the covariance-based SEM (CB-SEM) in that the primary objective of the former is to maximize the explained variance of the dependent constructs while the goal of the latter is to estimate a set of model parameters in a way to minimize the difference between the observed and the estimated variance-covariance matrices. The CB-SEM approach requires strict assumptions as the multivariate normality of data and a minimum sample size. When these assumptions are violated or when the goal of the analysis is the prediction, PLS-SEM could be the preferred method [9].

In the proposed approach, three latent variables are created according to the network obtained by addressing the first goal: the treatment variable, the mediator variable, and the disease variable (see Fig. 1).

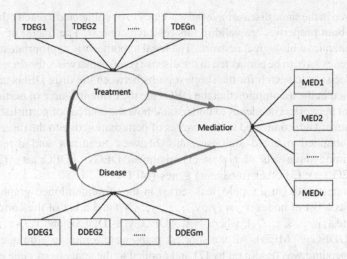

Fig. 1. Mediation model proposed from the connection of treatment genes to disease genes

2 Materials and Methods

The analyses were performed using the dataset GSE62117 [10] downloaded from the Gene Expression Omnibus (GEO) database. In particular, the dataset includes the samples obtained for patients affected by HIV and treated with abacavir (ABC) (10 samples), for untreated patients (31), and controls (15). Two comparisons were considered to obtain sets of DEGs: (1) ABC treatment vs. HIV samples; (2) HIV samples vs. controls. The comparisons were performed by using Significance Analysis of Microarray (SAM) [11]. SAM is a statistical technique that computes t-tests based on permutation. In the next step, Signaling Pathways Impact Analysis (SPIA) [12] was used to discover the KEGG pathways significantly associated each set of DEGs. SPIA combines enrichment analysis and perturbation analysis to determine the statistical significance of a pathway. Two p-values are computed: the pNDE and the pPERT. The pNDE value represents the probability of getting a number of DEGs on the given pathway at least large as the observed one. The pPERT is the probability of observing a total perturbation of the pathway larger than the one got by chance. The perturbation for a gene is defined in terms of the log-fold change of that gene and of those genes that are directly upstream. The so-obtained significant pathways were considered as a graph and joined together to get a unique graph determining how the treatment DEGs act on the disease DEGs at the network level.

To get the unique graph, the function mergeKEGGgraphs() from the R package KEGGgraph was used [13]. The function permits not only to merge graphs, but also to obtain the information of nodes and edges from KEGG. The geodesic distance $d_{geo}(y_i, y_j)$ and shortest paths between each pair of treatment-DEG y_i and disease-DEG y_j were computed. The use of shortest paths is motivated by two important properties: small world phenomena and local hypothesis. The first property affirms that in a biological network it is possible to connect any pair of nodes by shortest paths. The second property implies that genes involved in the same disease have the tendency to be connected to each other [2]. We believe that both properties are valid in this research context. The fusion of significant pathways generates a biological network. The local hypothesis is appropriate and logical as the drug genes have to be connected to the disease genes; otherwise, the drug would not work. The decision to search for the shortest paths between the drug DEGs and disease DEGs is rooted in the assumption that the DEGs are the evident source of perturbation in the biological network. The idea is to understand how the sources of perturbation due to the drug treatment are connected to the sources of perturbation due to the disease effects.

The so-obtained directed shortest paths between treatment and disease genes allowed defining three sets of nodes: (1) treatment DEGs (TDEG set); (2) disease DEGs (DDEG set); (3) other (mediator) genes (MED).

Each shortest path (in a topological sense) in the aforementioned graph could be represented as a list of nodes $Y_k = (y_i, y_{i+1},....., y_{j-1}, y_j)$ and a list of the corresponding edges $E_k = (e_{i(i+1)}, ..., e_{(j-1)j})$, where $y_i \in$ TDEG, $y_j \in$ DDEG, and $(y_{i+1}, ..., y_{j-1}) \in$ (TDEG $\cup$ DDEG $\cup$ MED). All the shortest paths were joined in a unique graph Y. A similar procedure was described by [7] and applied to the analysis of gene expression data [14, 15].

It was necessary to define these three new subsets as they do not correspond to the SAM lists. In fact, here only the genes connected involved in shortest paths are considered; the others are excluded. Furthermore, the mediation genes could be genes not differentially expressed but important they allow connecting the treatment DEGs to the disease DEG.

Finally, PLS-SEM was used to create the mediation model allowing the evaluation of the treatment effect on the disease (see Fig. 2). The PLS-SEM approach is a general method for estimating causal relationships in path models that involves latent constructs which are indirectly measured by various indicators. PLS path models are formally defined by two sets of equations: the inner (or structural) model and the outer (or measurement) model. The inner model specifies the relationships between the unobserved (or latent) variables, whereas the outer model specifies the relationship between a latent variable and its observed (or manifest) variables. In PLS, the outer relationships include two types of models: a formative one and a reflective one. The formative measurement model specifies cause–effect relationships between the manifest variables and the latent index (independent causes). The reflective measurement model involves paths from the latent construct to the manifest variables or dependent effects. The inner model can be represented by a linear equation:

$$\mathbf{Y} = \mathbf{YB} + \mathbf{Z} \tag{1}$$

where $\mathbf{Y}$ is the matrix of the latent variables, $\mathbf{B}$ is the matrix of coefficients, and $\mathbf{Z}$ is the matrix of error terms (assumed to be centered). The outer model, on the other hand, can be represented in the following way:

$$\mathbf{X_g} = \mathbf{y_g} w_G^T + \mathbf{F_g} \tag{2}$$

where X_g is the block matrix of observed variables associated with the latent variable y_g; w_G^T is the matrix of correlation coefficients between X_g and y_g, computed by least squares; and F_g is the error term.

In our analyses, all latent variables were defined to have mean zero and variance one. For the evaluation of the statistical significance of the path coefficients, a bootstrap approach was used [16]. The goodness of fit of the outer model was evaluated by using the Dillon-Goldstein rho, which focuses on the variance of the sum of observed variables associated with a latent variable. As a rule of thumb, a block was considered as unidimensional when the Dillon-Goldstein rho was larger than 0.7. For the evaluation of the inner model or the structural model, R^2 was computed. Values of R^2 were classified as follows:

1. Low: $R^2 < 0.30$;
2. Moderate: $0.30 < R^2 < 0.60$;
3. High: $R^2 > 0.60$

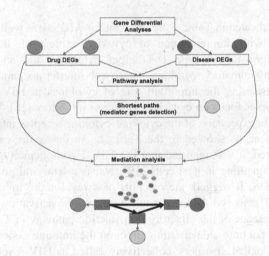

Fig. 2. The pipeline starts from the detection of DEGs for drug treatment and disease. Then by pathway analysis and shortest paths between drug DEGS and disease DEGs, the mediator genes are detected. The expression values of the three sets of genes are used to create the latent treatment, disease, and mediator variables for the final mediation analysis.

3 Results

The differential analysis yielded 7706 DEGs for the ABC treatment (based on comparing the treated-patient samples to the untreated-patient samples) and 177 DEGs for the HIV samples (based on comparing the untreated-patient samples to the control samples). Both lists were constructed by assuming FDR equal to 0.05 and a minimum fold change of 2. SPIA revealed three pathways for the treatment DEGs and two pathways for the disease DEGs (Table 1).

Table 1. Significant KEGG pathways for ABC treatment and AIDS datasets

Name	pSize	NDE	pNDE	pPERT	pGFdr	Status	DEGs Set
Cytokine-cytokine receptor interaction	243	122	0,000	0,001	0,000	Inhibited	ABC
Olfactory transduction	145	77	0,000	0,001	0,000	Inhibited	ABC
Calcium signaling pathway	176	95	0,000	0,996	0,022	Inhibited	ABC
Influenza A	105	5	0,003	0,001	0,002	Activated	AIDS
Herpes simplex infection	98	4	0,012	0,006	0,019	Activated	AIDS

pSize: number of genes in pathway; NDE: number of DEGs in pathways; pNDE: p-value enrichment analysis; pPERT: p-value on total perturbation; pGFdr: adjusted FDR for global p-value; Status: direction of perturbation; DEGs Set: the DEGs set associated with the pathway.

The pathways indicated in Table 1 are relevant for AIDS. It is well known that HIV infection interferes with the regulation of cytokine expression. In fact, a general decrease in the expression of type 1 T-helper cytokines and an increase in the expression of proinflammatory cytokines, antiviral interferons, and TGF-beta was observed [17]. Considering the important role of cytokines in HIV infection, many immune-based therapies focus on cytokines and their inhibitors [18]. This explains the observed inhibition in expression of the cytokine-cytokine receptor interaction pathway by the ABC treatment, as showed in the Table 1. As far as the calcium signalling pathway is considered, it was reported that the HIV-1 pathogenicity factor Nef modulates the calcium signalling in host cells [19]. Nef is a lentiviral protein involved in pathogenesis of AIDS. It triggers the calcium pathway for the induction of nuclear factor of activated T cells (NFAT) that could promote the activation of HIV-infected T cells [20]. The presence of the olfactory transduction pathway in Table 1 is also not surprising. HIV has not only a devastating effect on the immune system, but it can also cause several neurological disorders, collectively called as HIV-associated neurocognitive disorders (HAND) [21]. Among the neurological disorders, deficits in olfaction are common and the severity and the onset of the disease can be established by olfactory test [22, 23]. Finally, influenza A and herpes simplex infection pathways are involved in the activation of immune system, as well as HIV. The infection with the herpes simplex infection can increase the risk of HIV acquisition among men and women [24].

The pathways from Table 1 were merged and transformed in a unique graph constituted by 675 nodes and 1855 connections. There were 13 shortest paths between the ABC treatment DEGs and the AIDS DEGs, for a total of 15 nodes and 15 connections (see Table 2 and Fig. 3).

There are nine ABC DEGs, four AIDS DEGs, and two mediator genes. This means that only for a few ABC DEGs it was possible to find a path that connected them to the disease genes. The fusion of all shortest pasts constitutes the treatment module. Most of the genes included in Table 2, such as the arrestins, inhibins, activins, or the

Table 2. List of genes in the detected shortest paths.

Entrez	Name	Description
254973	or1l4	olfactory receptor, family 1, subfamily L, member 4
92	acvr2a	activin A receptor, type IIA
90	acvr1	activin A receptor, type I
8200	gdf5	growth differentiation factor 5
650	bmp2	bone morphogenetic protein 2
157	adrbk2	adrenergic beta receptor kinase 2
655	bmp7	bone morphogenetic protein 7
26539	or10h1	olfactory receptor, family 10, subfamily H, member 1
3626	inhbc	inhibin, beta C
3624	inhba	inhibin, beta A
409	arrb2	arrestin, beta 2
442194	or10c1	olfactory receptor, family 10, subfamily C, member 1
83729	inhbe	inhibin, beta E
269	amhr2	anti-Mullerian hormone receptor, type II
268	amh	anti-Mullerian hormone

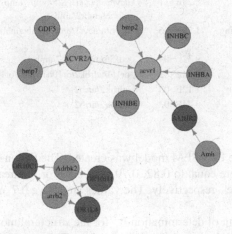

Fig. 3. The graph constructed from the shortest paths between the ABC treatment DEGs and the AIDS DEGs. Yellow nodes are the mediator genes, the red nodes are the AIDS DEGs, and the green nodes the ABC DEGs.

beta-adrenergic kinase gene, are known to be important for AIDS. Arrestins are fundamental adaptors connecting receptors to cell trafficking machinery. β-arrestin 2 could play a fundamental role in apoptosis and cell proliferation [25]. Activin A is released during acute systemic inflammation as a part of the circulatory cytokine cascade. It can play a role in pro- and anti-inflammatory actions on key genes for the inflammatory response as TNF-α, IL-1β and IL-6 [26]. Inhibins are negative regulators of activin activity [27]. Regarding the beta-adrenergic kinase gene, it was discovered that HIV coats protein gp120 interfers with the beta-adrenergic regulation of astrocytes and microglia. Hence, it may alter astroglial reactivity and upset the delicate cytokine network responsible for the defense against viral and opportunistic infections [28]. The anti-Mullerian hormone receptors play an important role associated to the TGF-β signalling pathway. The hormone is a ligand of the TGF-β [29].

A KEGG enrichment analysis of the genes in the two modules (treatment and disease) revealed that 10 of the 15 genes are a part of the two most significant pathways, TGF beta signalling pathway and the cytokine-cytokine receptor interaction (see Table 3). The HIV protein Tat has immunosuppressive effects as transforming growth factor-beta (TGF beta) and inhibiting the proliferation of lymphocyte. It is evident that HIV perturbs the TGF-beta signalling pathway [30].

Table 3. KEGG enrichment analysis of the genes present in the graph obtained by merging the shortest paths between drug DEGs and disease DEGs.

Description	p.adjust	geneID	Count
TGF-beta signaling pathway	1.30E−15	acvr2a/acvr1/gdf5/bmp2/bmp7/inhbc/inhba/ inhbe/amhr2/amh	10
Cytokine-cytokine receptor interaction	8.28E−11	acvr2a/acvr1/gdf5/bmp2/bmp7/inhbc/inhba/ inhbe/amhr2/amh	10
Signaling pathways regulating pluripotency of stem cells	1.26E−06	acvr2a/acvr1/bmp2/inhbc/inhba/inhbe	6
Hippo signaling pathway	9.32E−04	gdf5/bmp2/bmp7/amh	4
Olfactory transduction	3.94E−03	or1l4/adrbk2/or10h1/arrb2/or10c1	5
Hedgehog signaling pathway	1.12E−02	adrbk2/arrb2	2
Morphine addiction	3.44E−02	adrbk2/arrb2	2

In the next step, the PLS-SEM model was created. The Dillon-Goldstein rho values for the outer model were equal to 0.82, 0.79, and 0.83 for the treatment, the mediator, and the disease variables, respectively. The values are above 0.7 and can be considered acceptable.

The mean coefficient of determination R^2 for the structural model was equal to 0.5. Also, this value could be considered acceptable. The outer model connected the treatment variable to the disease variable directly or indirectly by the mediator variable. The bootstrap approach based on the fitted PLS-SEM allowed evaluating the statistical significance of the connection of the inner model. In Fig. 4, the estimated values of the coefficients of the mediation model are illustrated. All the coefficients were statistically significant at the 5% significance level, as the 95% confidence intervals did not include zero. The direct effect was estimated to be equal to 0.78 (coefficient between Treatment and Disease variables), the indirect effect to 0.08 (multiplication of the coefficients for Treatment vs. Mediator, −0.48 and for Mediator vs. Disease, −0.17), and the total effect to 0.86 (sum of the direct and indirect effects).

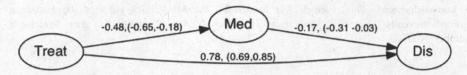

Fig. 4. The evaluation of the mediation analysis by PLS-SEM

4 Conclusion

In this paper, we have proposed and illustrated a new procedure, based on gene-expression data, to evaluate whether the effect of a treatment on a disease is statistically significant or not. We propose a mediation analysis based on three fundamental entities, gene-expression data, network information, and PLS-SEM. The first element allows determining the DEGs associated with the treatment and the disease. The network information is important to understand how the treatment DEGs are biologically connected to the disease DEGs. Direct shortest paths, between each pair of treatment DEG and disease DEG are searched for. The shortest paths were chosen according to the local hypothesis and small world phenomena principles. If no path is found between treatment DEGs and disease DEGs, the local hypothesis can be rejected. The reasons could be different: low quality data, bad choice of DEGs, or lack of effect of the treatment on the disease. The proposed approach is a downstream analysis; therefore, it assumes that no issues were detected in the previous analyses. The path can also include (mediator) genes not present in the set of treatment or disease DEGs.

The mediation model is created by PLS-SEM. To our knowledge, it is the first time that a mediation analysis is generated from gene expression data and gene network information. The method was applied to the GSE62117 dataset to evaluate the effect of the ABC treatment on AIDS. After the detection of DEGs and significant pathways for each set, the method indicated statistically significant direct and indirect effects of the treatment on the disease. The direct effect was more pronounced, as captured by the value of the corresponding coefficient equal to 0.78, compared to 0.08 for the indirect effect.

The procedure could offer a useful supporting tool when investigating the question whether, from a gene expression point of view, a treatment has got a significant effect on a disease and, if so, which molecular mechanisms are involved in the effect.

It is important to keep on mind that the proposed procedure is a multi-step one. To reduce uncertainty in the analysis, we recommend to verify the biological validity of the results for each of the involved steps. For example, the shortest paths and the generation of the three latent variables strongly depend on the differential and pathway analyses. The mediation analysis assumes the causal sufficiency of the mediation model. This means that there are no hidden confounders. It is, admittedly, a strong assumption. However, the goal is to obtain information about the relationship between treatment and disease based only on gene expression data. To make the assumption more likely to be fulfilled, one could consider integrating other "omics" data, as epigenomics, in the proposed approach. This is a topic for further research.

Acknowledgement. This research was funded by the MIMOmics grant of the European Union's Seventh Framework Programme (FP7-Health-F5-2012) under the grant agreement number 305280.

References

1. Clarke, P.A., et al.: Gene expression microarray analysis in cancer biology, pharmacology, and drug development: progress and potential. Biochem. Pharmacol. **62**(10), 1311–1336 (2001)
2. Barabási, A.L., Gulbahce, N., Loscalzo, J.: Network medicine: a network-based approach to human disease. Nat. Rev. Genet. **12**(1), 56–68 (2011)
3. Baron, R.M., Kenny, D.A.: The moderator-mediator variable distinction in social psychological research – conceptual, strategic, and statistical considerations. J. Pers. Soc. Psychol. **51**(6), 1173–1182 (1986)
4. Gunzler, D., Chen, T., Wu, P., Zhang, H.: Introduction to mediation analysis with structural equation modeling. Shanghai Arch. Psychiatry **25**(6), 390 (2013). doi:10.3969/j.issn/1002-0829.2013.06.009
5. MacKinnon, D.: Introduction to Statistical Mediation Analysis. Lawrence Erlbaum Associates, New York (2008). doi:10.4324/9780203809556
6. Kanehisa, M., Goto, S.: KEGG: kyoto encyclopedia of genes and genomes. Nucleic Acids Res. **28**(1), 27–30 (2000)
7. Pepe, D., Grassi, M.: Investigating perturbed pathway modules from gene expression data via structural equation models. BMC Bioinform. **15**(1), 1–15 (2014)
8. Hair Jr., J.F., et al.: A Primer on Partial Least Squares Structural Equation Modeling (PLS-SEM). Sage Publications, Los Angeles (2013)
9. Hair, J.F., Ringle, C.M., Sarstedt, M.: PLS-SEM: indeed a silver bullet. J. Market. Theory Pract. **19**(2), 139–152 (2011)
10. Mohsen, S., et al.: Differential adipose tissue gene expression profiles in abacavir treated patients that may contribute to the understanding of cardiovascular risk: a microarray study. PLoS ONE **10**(1), e0117164 (2015)
11. Tusher, V.G., Tibshirani, R., Chu, G.: Significance analysis of microarrays applied to the ionizing radiation response. Proc. Natl. Acad. Sci. **98**(9), 5116–5121 (2001)
12. Tarca, A.L., et al.: A novel signaling pathway impact analysis. Bioinformatics **25**(1), 75–82 (2009)
13. Zhang, J.D., Wiemann, S.: KEGGgraph: a graph approach to KEGG PATHWAY in R and bioconductor. Bioinformatics **25**(11), 1470–1471 (2009)
14. Pepe, D., Do, J.H.: Comparison of perturbed pathways in two different cell models for parkinson's disease with structural equation model. J. Comput. Biol. **23**(2), 90–101 (2016)
15. Pepe, D., Do, J.H.: Estimation of dysregulated pathway regions in MPP + treated human neuroblastoma SH-EP cells with structural equation model. BioChip J. **9**(2), 131–138 (2015)
16. Monecke, A., Leisch, F.: semPLS: structural equation modeling using partial least squares. J. Stat. Softw. **48**(3), 1–32 (2012)
17. Valdez, H., Lederman, M.M.: Cytokines and cytokine therapies in HIV infection. AIDS Clin. Rev. 187–228 (1996)
18. Kim, J.J., Simbiri, K.A., Sin, J.I., Dang, K., Oh, J., Dentchev, T., Lee, D., Nottingham, L.K., Chalian, A.A., Mccallus, D., Ciccarelli, R.: Cytokine molecular adjuvants modulate immune responses induced by DNA vaccine constructs for HIV-1 and SIV. J. Interferon Cytokine Res. **19**(1), 77–84 (1999)

19. Manninen, A., Saksela, K.: HIV-1 NEF interacts with inositol trisphosphate receptor to activate calcium signaling in T cells. J. Exp. Med. **195**(8), 1023–1032 (2002)
20. Manninen, A., Renkema, G.H., Saksela, K.: Synergistic activation of NFAT by HIV-1 NEF and the Ras/MAPK pathway. J. Biol. Chem. **275**(22), 16513–16517 (2000)
21. Lindl, K.A., Marks, D.R., Kolson, D.L., Jordan-Sciutto, K.L.: HIV-associated neurocognitive disorder: pathogenesis and therapeutic opportunities. J. Neuroimmune Pharmacol. **5** (3), 294–309 (2010)
22. Vance, D.E.: Olfactory and psychomotor symptoms in HIV and aging: potential precursors to cognitive loss. Med. Sci. Monitor **13**, SC1–SC3 (2007)
23. Zucco, G.M., Ingegneri, G.: Olfactory deficits in HIV-infected patients with and without AIDS dementia complex. Physiol. Behav. **80**, 669–674 (2004)
24. Freeman, E.E., Weiss, H.A., Glynn, J.R., Cross, P.L., Whitworth, J.A., Hayes, R.J.: Herpes simplex virus 2 infection increases HIV acquisition in men and women: systematic review and meta-analysis of longitudinal studies. Aids **20**(1), 73–83 (2006)
25. Moorman, J., Zhang, Y., Liu, B., LeSage, G., Chen, Y., Stuart, C., Prayther, D., Yin, D.: HIV-1 gp120 primes lymphocytes for opioid-induced, β-arrestin 2-dependent apoptosis. Biochimica et Biophysica Acta (BBA)-Molecular Cell Res. **1793**(8), 1366–1371 (2009)
26. Jones, K.L., De Kretser, D.M., Patella, S., Phillips, D.J.: Activin A and follistatin in systemic inflammation. Mol. Cell. Endocrinol. **225**(1), 119–125 (2004)
27. Phillips, D.J., de Kretser, D.M., Hedger, M.P.: Activins and related proteins in inflammation: not just interested bystanders. Cytokine Growth Factor Rev. **20**(2), 153–164 (2009)
28. Levi, G., et al.: Human immunodeficiency virus coat protein gp120 inhibits the beta-adrenergic regulation of astroglial and microglial functions. Proc. Natl. Acad. Sci. **90** (4), 1541–1545 (1993)
29. Di Clemente, N., Jamin, S.P., Lugovskoy, A., Carmillo, P., Ehrenfels, C., Picard, J.Y., Whitty, A., Josso, N., Pepinsky, R.B., Cate, R.L.: Processing of anti-mullerian hormone regulates receptor activation by a mechanism distinct from TGF-β. Mol. Endocrinol. **24**(11), 2193–2206 (2010)
30. Lotz, M., Clark-Lewis, I., Ganu, V.: HIV-1 transactivator protein Tat induces proliferation and TGF beta expression in human articular chondrocytes. J. Cell Biol. **124**(3), 365–371 (1994)

A Novel Algorithm for CpG Island Detection in Human Genome Based on Clustering and Chaotic Particle Swarm Optimization

Abdelbasset Boukelia[✉], Zakaria Benmounah, Mohamed Batouche,
Bouchera Maati, and Ikram Nekkache

Computer Science Department, Faculty of NTIC,
University of Constantine 2 - Abdelhamid Mehri, Constantine, Algeria
{abdelbasset.boukelia,zakaria.benmounah,
mohamed.batouche}@univ-constantine2.dz

Abstract. CpG islands provide a major role in the genome and are used for prediction of promoter regions. They are abnormally methylated in cancer cells and can be used as tumor markers. However, current techniques for identifying CpG islands suffer from various drawbacks. In this paper, we propose a novel algorithm to detect CpG islands by combining clustering techniques with complementary chaotic particle swarm optimization. Clustering techniques are used to find the locations of potential CpG island candidates in the genome while Complementary Chaotic PSO is used to find the best location of a CpG island in a cluster candidate without being trapped in local optimum solution. This combination can successfully overcome the drawbacks of each method while maintaining their advantages. The proposed method called 3C-PSO provides a high sensitivity detection of CpG islands in the human genome. To evaluate its performance, we used six sequences from NCBI, and five measures of performance: sensitivity (SN), specificity (SP), accuracy (ACC), performance coefficient (PC), and correlation coefficient (CC). We compared our approach to the existing methods of CpG islands detection in the human genome. The obtained results have shown that 3C-PSO competes with and even outperforms these methods.

Keywords: CpG island · Genome · Clustering · Chaotic map · Particle swarm optimization

1 Introduction

Most researchers equate epigenetics with the study of chromatin-marking systems (DNA methylation, histone modification, and chromatin remodeling) [1]. DNA methylation in the mammalian genome refers to the methylation of cytosine within a CpG island [2][3]. CpG islands are highly enriched with CG nucleotides regions. These CpG sites are short line segments where a cytosine nucleotide is

© Springer International Publishing AG 2017
A. Bracciali et al. (Eds.): CIBB 2016, LNBI 10477, pp. 70–81, 2017.
DOI: 10.1007/978-3-319-67834-4_6

followed by a Guanine nucleotide in the direction 5' to 3' [3]. A CpG is an abbreviation for 5'-C-phosphate G-3' which is a cytosine and guanine separated by a single phosphate [4]. The cytosine in CpG dinucleotides may be methylated to form a methylcytosine base mC or unmethylated to return back to a cytosine base "C". CpG islands play an important role in DNA methylation [5]. Most CpG islands are sites of transcription initiation. The methylation of a gene promoter seems to be closely associated with the inactivation or inhibition of the transcription of this gene [6], but nonmethylation of these promoter sites (unmethylated CpG islands) of a non-transcribed gene can induce its transcription [7].

Hypermethylation of CpG islands located in the promoter regions of tumor suppressor genes is now firmly established as an important mechanism for gene inactivation [1,8]. CpG island hypermethylation has been described in almost every tumor type [1]. The development of CpG island hypermethylation profiles for every form of human tumors has yielded valuable pilot clinical data in monitoring and treating cancer patients based on our knowledge of DNA methylation [9].

The most used procedure to locate the potential CpG islands in the genome is by looking at regions of the DNA with at least 200 nucleotides in length, where the GC percentage is at least 50% and an observed-to-expected CpG ratio is above 60% [10]. In this paper, a novel method for CpG island detection is proposed. It consists mainly of two stages: A clustering stage which allows filtering the genome in order to extract the potential CpG island candidates, and a second stage which refines the island candidates by locating the best CpG islands. The rest of the paper is organized as follows. In Sect. 2, we present recent works related to CpG island detection and prediction. Section 3 is dedicated to the description of the proposed approach for CpG island detection. In Sect. 4, we present the experimental results. Finally, conclusions and future work are drawn.

2   Related Work

There are several computer programs written in a variety of languages which locate CpG islands in DNA sequences. CpGcluster [11] is used to predict statistically significant clusters of CpG dinucleotides by calculating the distance between CpG sites and determining a threshold to create CpG island clusters.

The CPSORL method [12] combines complementary particle swarm optimization (CPSO) with the reinforcement learning (RL) to predict CpG islands in the human genome. This method uses GGF criteria [10] as guidelines to identify CpG islands. CPSORL is composed of two major steps. First, the sequence is divided into segments, and then the PSO algorithm is used to locate CpG Island by updating iteratively the search of optimal results and identifying the best particles in the swarm population using the GGF criteria as fitness function. In the standard PSO, the particles could be trapped into a local optimum due to the premature convergence of particles. The complementary strategy aims to assist the particle search ability which helps the particle deviating in a local optimum by moving their position to a new region in the search space. The Chaotic

PSO method [13] is based on PSO with adaptive inertia weight factor (AIWF) and chaotic local search. The chaos dynamic is used for exploration by updating particle swarm using a Gaussian discrete chaotic map [14]. The ClusterPSO method [15] combines CpGCluster research method and a PSO algorithm. The CpGCluster method is used to construct CpG island candidate clusters. These clusters are used as region candidates for PSO algorithm to find the best CpG islands.

In the proposed approach, we have presented a novel method for CpG islands detection within the human genome using three strategies namely Clustering, Complementary PSO, and chaotic theory (3C-PSO). We implemented a different combination of PSO with an assessment of the relative importance of genes by examining the fitness function of the genes in the cluster limit. The 3C-PSO provides a higher prediction performance and combination ability than other algorithms as the statistics analysis show.

3   The Proposed Approach for CpG Island Detection

The proposed method for CpG island detection combines a clustering technique with an optimization technique. The clustering is used to find potential CpG islands by using distances between CpG sites and thresholds while the optimization technique is used to refine the search and yields the best CpG islands included inside the CpG island cluster candidates. The optimization technique uses the complementary chaotic PSO for each CpG island cluster candidate. This algorithm can be described as follows.

– **Search of CGI candidate using the Clustering.** Two important properties are necessary to generate clusters:
    • The positions of GC sites.
    • The distance between GC sites.
    In this phase, the Clustering proceeds as follows:
    1. Record the position 'C' of GC sites by scanning the DNA sequence in the direction 3' to 5' to collect the positions $C=(c1,c2,...cn)$ where 'n' represents the number of GC sites.
    2. Calculate the distance between adjacent GC sites which is estimated using the following distance measure:

$$D_i = c_{i+1} - c_i - 1 \tag{1}$$

    the minimum distance between adjacent sites '$CGCG$' equals 1.
    3. Sort the distance list without eliminating any repeated distance in order to find out the threshold 'd_f' in the position 65th of the list;
    4. Collect the positions using the threshold to generate clusters. When a distance between neighboring GC sites is smaller than the threshold d_f, the two neighboring sites belong to the same cluster otherwise create a new cluster. Repeat this step to generate all possible clusters.

5. After the determination of all clusters, the **p-value** of each cluster is calculated for estimating the probability to discover a CpG cluster in a random sequence. The negative binomial distribution is calculated by the cumulative density function at point n_f of the CpG cluster, and is taken as the *p-value*.

$$P_{(N,p)}^{Cum}(x <= n_f) = \sum_{x=0}^{n_f} \binom{(x-(N+1)-1)}{((N-1)-1)} * p^{N-1} * (1-p)^x \qquad (2)$$

$$n_f = L - 2 * N \qquad (3)$$

$$p = N_s/N_{is} \qquad (4)$$

N is the number of CpGs in the cluster, n_f is the number of independent non-CpGs in the cluster, L is the length of the cluster, and p is the probability of success discovering a CpG. N_s is the number of CpGs and N_{is} is the number of independent dinucleotides in the DNA sequence. This phase examines statistically significant CpG clusters and assumes that all CpG islands are included in these clusters. If the *p-value* of a cluster is smaller than the threshold value, then the cluster is accepted, otherwise the cluster is rejected.

– **The optimization of CGIs candidates with the Complementary Chaotic PSO.** this phase optimizes the CGI candidates by following these steps:

1. Extend the cluster candidates with 200 dinucleotides (100 dinucleotides in both sides of the cluster).
2. Initialization of the swarm randomly:
   - Set the position and velocity for each particle randomly, the position encoding is given by:

$$P_i = (F_{si}, F_{li}) \qquad (5)$$

Where F_{si} is the predicted start position of the particle P_i and F_{li} is its predicted length in the cluster. The F_{si} and F_{li} are initialized by using Eqs. 6 and 7 and the limits of each cluster

$$F_{si}^{initial} = rand * (end_{sequence} - 200 - start_{sequence}) + start_{sequence} \qquad (6)$$

$$F_{li}^{initial} = rand * (end_{sequence} - 200 - F_{si}^{initial}) + 200 \qquad (7)$$

   - Set the initial value of the chaos as follows:

$$chaos_0 = rand * 2 - 1 \qquad (8)$$

3. The fitness evaluation must respect the GGF criteria:

$$GC_{content} \geq 0.50, Obs/Exp_{ratio} \geq 0.6, CGI_{length} \geq 200bp$$

The fitness functions of the length, the GC content and the O/E ratio are defined in Eqs. 9–11, respectively. In addition, Eq. 12 is used to calculate the fitness value of each particle [12, 15]. Note that the fitness values must be between 0 and 1 to adjust the function result:

$$CpG_{length}(P_i) = \frac{F_{li}}{L} \tag{9}$$

$$GC(P_i) = \frac{\#C + \#G}{F_{li}} \tag{10}$$

$$Obs/_{Exp}(P_i) = \frac{\#CpG * F_{li}}{\#C * \#G} \tag{11}$$

$$Fitness(P_i) = CpG_{length}(P_i) + GC(P_i) + Obs/_{Exp}(P_i) \tag{12}$$

Where #C, #G, and #CpG are respectively the numbers of nucleotide cytosine (C), guanine (G) and the number of CpG sites in the predicted CpG island region at Pi. L is the number of nucleotide in the considered cluster (extended CpG island candidate).

4. Update the *pbesti* (best solution for particle i so far) and *gbest* (global best solution so far) for all particles.
5. Update the Chaotic Gauss map value to enrich the searching behavior. The logical equation is defined as follows:

$$chaos_{n+1} = \exp\left(-alpha * chaos_n^2\right) + beta \tag{13}$$

where *alpha* and *beta* are set to 4.90 and –0.58 respectively. The parameter settings are based on bifurcation diagram as shown on Fig. 1 so that to get chaotic behavior [16]. With alpha set to 4.90, beta should be set to values in the range [–0.58, –0.4]. Fig. 2 shows some plots of the Gauss map with this parameter setting.

6. Update the velocity and position of Pi according to Eqs. 14 and 15, based on *pbesti* and global best (*gbest*)

$$V_{(i,j)}^{new} = w * V_{(i,j)}^{old} + chaos * c_1 * (pbest_i - x_{(i,j)}^{old})$$
$$+ chaos * c2 * (gbest_i - x_{(i,j)}^{old}) \tag{14}$$

$$x_{(i,j)}^{new} = x_{(i,j)}^{old} + V_{(i,j)}^{new} \tag{15}$$

The parameters $c1$ and $c2$ are called acceleration coefficients (usually $c1 = c2$). Chaos is a function based on the results of the chaotic map with values between –1.0 and 1.0. $V_{(i,j)}^{new}$ and $V_{(i,j)}^{old}$ denote respectively, the velocities of new and old particle. $x_{(i,j)}^{old}$ is the current particle position, and $x_{(i,j)}^{new}$ is the updated particle position. w is called inertia weight and decreases linearly from 0.9 to 0.4 throughout the search process. The weight equation can be written as:

$$w_i = (w_{max} - w_{min}) * \frac{Iteration_{max} - Iteration_i}{Iteration_{max}} + w_{min} \tag{16}$$

Where w_{max} is equal to 0.9 and w_{min} is equal to 0.4. $Iteration_{max}$ is the maximum number of allowed iterations and $Iteration_i$ corresponds to i^{th} iteration.

7. If the fitness of *gbest* does not change during five consecutive iterations, a particle is considered trapped in a local optimum. To avoid that, the treatment process is moved to a new region of the search space. We select randomly 50% of the population which will undergo a change. The positions of the selected particles are used to generate complementary particles. The position of a trapped particle is changed based on the following complementary rules:

$$x_{id}^{Complement} = (x_{min} + x_{max}) - x_{id}^{selected} \tag{17}$$

Where $x_{id}^{selected}$ is the position of a randomly selected particle, and $x_{id}^{Complement}$ is the position of its complementary particle. x_{max} and x_{min} denote the maximum and minimum limits of the solution space respectively.

More formally, the whole algorithm is described on Fig. 3.

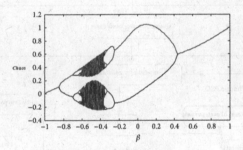

Fig. 1. Bifurcation diagram of the gauss map with *alpha* = 4.90 and *beta* in the range −1.0 to +1.0.

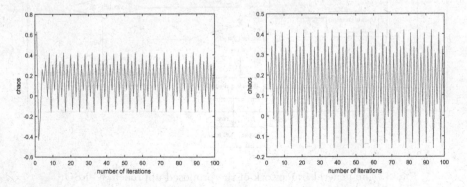

Fig. 2. Some plots of the gauss map with *alpha* = 4.90 and *beta* = −0.58.

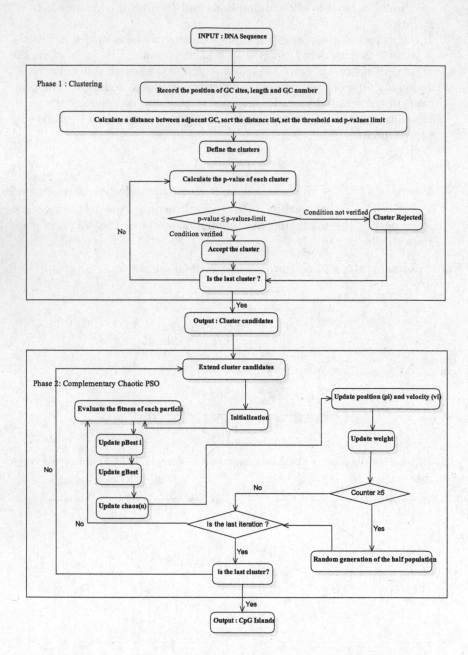

Fig. 3. The general framework of the proposed approach 3C-PSO.

4   Experimental Results

3C-PSO shows a high sensitivity detection of CpG islands in the human genome thanks to the used techniques. In the clustering technique, a distance threshold parameter was set to 65th position and the $p - value$ to 0.01. In the complementary chaotic PSO, the population size was set to 300 particles. The number of iterations is set to 100 and $c1 = c2 = 2$. The GGF parameters used to define CpG islands were: the length set to minlength $= 200$ bp, GC content set to 0.5, O/E ratio set to 0.6, and the gap between adjacent islands set to 100 bp. We used five common criteria to determine the prediction accuracy, namely the sensitivity (SN), specificity (SP), accuracy (ACC), performance coefficient (PC) and correlation coefficient (CC).

In this study, we calculated the five prediction performances which are defined as follows

$$SN = \frac{TP}{TP + FN} \tag{18}$$

$$SP = \frac{TN}{TN + FP} \tag{19}$$

$$ACC = \frac{TP + TN}{TP + TN + FP + FN} \tag{20}$$

$$CC = \frac{TP * TN - FP * FN}{\sqrt{(TP + FN) * (TP + FP) * (TN + FP) * (TN * FN)}} \tag{21}$$

$$PC = \frac{TP}{TP + FN + FP} \tag{22}$$

where TP, TN, FN and FP refer respectively to true positive, true negative, false negative and false positive.

We compared our approach with more than five other methods reported in the literature like CpGplot, CpGCluster, PSORL.

Tables 1 and 2 show that the SP of the proposed method was highest on the NT 113952.1 (99.51%), NT 113955.2 (100.0%), NT 113958.2 (100.0%), NT 113953.1(99.98%), NT 113954.1 (99.399%) and NT 028395.3 (100%) and the ACC measure is high in all the six sequences than other methods. The PC is also high in NT 113958.2 and NT 113953.1.

Tables 3 and 4 show the number of CpG islands identified by CPSORL and 3C-PSO. It is shown that the proposed approach provides better results than other methods.

Table 1. Prediction performance of detecting CpG islands with other methods (1)

Contig.	Performance	Methods			
		CpGPlot	CpGcluster	ClusterPSO	3C-PSO
NT 113952.1	SN	56.43	50.46	95.98	92.46
	SP	100	99.95	99.47	100
	ACC	98.09	97.78	99.32	99.51
	PC	56.42	49.92	86.16	74.94
	CC	74.38	69.41	92.28	85.67
NT 113955.2	SN	47.19	67.15	94.67	84.78
	SP	100	99.72	99.51	100
	ACC	98.08	98.54	99.33	99.98
	PC	47.14	62.47	83.81	84.78
	CC	67.94	77.03	90.92	93.94
NT 113958.2	SN	51.29	27.16	88.56	88.84
	SP	99.99	99.94	99.1	100
	ACC	96.9	95.32	98.43	99.9
	PC	51.24	26.92	78.2	88.41
	CC	70.38	49.96	86.93	93.98
NT 113953.1	SN	22.8	57.32	82.74	96.72
	SP	100	99.74	99.47	99.98
	ACC	97.76	98.51	98.99	99.97
	PC	22.8	52.74	70.39	93.09
	CC	47.21	69.89	82.09	96.37
NT 113954.1	SN	31.24	29.86	78.02	84.65
	SP	100	99.46	98.23	100
	ACC	97.47	96.9	97.48	99.32
	PC	31.24	26.19	53.34	50.81
	CC	55.17	43.81	68.72	71.18
NT 028395.3 Length = 647850	SN	27.11	44.89	81.53	72.19
	SN	27.11	44.89	81.53	72.19
	SP	100	99.47	99.24	100
	ACC	97.98	97.53	98.6	99.94
	PC	27.1	39.26	67.53	72.2

Table 2. Prediction performance of detecting CpG islands with other methods (2)

Contig.	Performance	Methods				
		PSORL	CPSO	CPSORL	ChaoticPSORL	3C-PSO
NT 113952.1	SN	75.58	77.43	84.88	86.99	92.46
	SP	99.02	99.58	99.05	99.43	100
	ACC	97.99	98.61	98.43	98.88	99.51
	PC	62.27	70.91	70.34	77.31	74.94
	CC	75.71	82.49	81.8	86.62	85.67
NT 113955.2	SN	59.63	77.8	87.38	90.12	84.78
	SP	99.88	99.5	99.61	99.77	100
	ACC	98.42	98.71	99.16	99.43	99.98
	PC	57.74	68.67	79.08	85.08	84.78
	CC	74.51	80.85	87.89	91.66	93.94
NT 113958.2	SN	81.65	81.08	84.11	87.44	88.84
	SP	97.9	98.17	98.34	98.27	100
	ACC	96.87	97.08	97.43	97.58	99.9
	PC	62.33	63.8	67.51	69.66	88.41
	CC	75.28	76.41	79.31	80.99	93.98
NT 113953.1	SN	64.8	70.53	75.65	76.33	96.72
	SP	99.23	99.22	99.13	99.25	99.98
	ACC	98.23	98.38	98.45	98.59	99.97
	PC	51.59	55.91	58.57	61.14	93.09
	CC	67.25	70.9	73.1	75.15	96.37
NT 113954.1	SN	63.58	70.54	77.68	78.21	84.65
	SP	98.13	98.34	98.23	98.51	100
	ACC	96.86	97.32	97.48	97.76	99.32
	PC	42.74	49.22	53.15	56.36	50.81
	CC	58.36	64.72	68.53	71.17	71.18
NT 028395.3 Length = 647850	SN	72.79	72.52	77.02	71.43	72.19
	SN	72.79	72.52	77.02	71.43	72.19
	SP	98.99	99.18	98.9	99.06	100
	ACC	98.06	98.24	98.12	98.08	99.94
	PC	57.17	59.36	59.25	56.92	72.2

Table 3. Number of CpG islands located in gene regions identified with CPSORL/3CPSO.

Chr	Contig	Number CpG islands detected		Number of true islands	
		CPSORL	3C-PSO	CPSORL	3C-PSO
21	NT 113952.1	12	15	1(3)	12
21	NT 113955.2	15	15	2(3)	11
21	NT 113958.2	19	31	2(3)	23
21	NT 113953.1	8	10	1(1)	8
21	NT 113954.1	10	15	1(1)	10
22	NT 028395.3	38	46	10(15)	29

Table 4. Length and GC% average of CpG islands located in gene regions with CPSORL/3CPSO

Chr	Contig	GC% (Average)		CpG island length	
		CPSORL	3C-PSO	CPSORL	3C-PSO
21	NT 113952.1	0.51886	0.664308	8537	6615
21	NT 113955.2	0.5	0.6203543	10023	11075
21	NT 113958.2	0.5	0.662772	14470	10604
21	NT 113953.1	0.5	0. 595321	3998	2974
21	NT 113954.1	0.5037	0.620093	6174	3451
22	NT 028395.3	0.5	0.6281725	24649	14472

5 Conclusion and Future Work

In this paper, we have presented a hybrid method called 3C-PSO (Clustering and Complementary Chaotic PSO) to detect and predict CpG islands in the human genome. The proposed method combines two main techniques: a clustering technique and an optimization technique. This combination allows achieving CpG island detection in the human genome with high accuracy. Indeed, the clustering technique is used to filter the human genome in order to obtain good quality potential CpG island cluster candidates. These candidates are then refined by using an optimization technique namely Complementary Chaotic PSO in order to find out the right CpG islands present in the human genome. Experimental results have shown that the proposed method competes and even outperforms existing methods for CpG island detection. As future work, we expect using parallel and distributed computing (Hadoop MapReduce, Apache Spark) in order to improve the performance of the proposed method.

Acknowledgments. Authors would like to thank the National Center for Biotechnology Research (CRBt) Constantine http://www.crbt.dz for its partial support to this work.

References

1. Herman, J.G., Graff, J.R., Myöhänen, S.B.D.N., Nelkin, B.D., Baylin, S.B.: Methylation-specific pcr: a novel pcr assay for methylation status of cpg islands. Proc. Natl. Acad. Sci. **93**(18), 9821–9826 (1996)
2. Viré, E., et al.: The polycomb group protein ezh2 directly controls dna methylation. Nature **439**(7078), 871–874 (2005)
3. Bird, A.: Dna methylation patterns and epigenetic memory. Genes Dev. **16**(1), 6–21 (2002)
4. Jones, P.A., Takai, D.: The role of dna methylation in mammalian epigenetics. Science **293**(5532), 1068–1070 (2001)

5. Gonzalez-Zulueta, M., Bender, C.M., Yang, A.S., Nguyen, T.D., Beart, R.W., Van Tornout, J.M., Jones, P.A.: Methylation of the 5' cpg island of the p16/cdkn2 tumor suppressor gene in normal and transformed human tissues correlates with gene silencing. Cancer Res. **55**(20), 4531–4535 (1995)
6. Keshet, I., Lieman-Hurwitz, J., Cedar, H.: Dna methylation affects the formation of active chromatin. Cell **44**(4), 535–543 (1986)
7. Cotton, A.M., Lam, L., Affleck, J.G., Wilson, I.M., Peñaherrera, M.S., McFadden, D.E., Kobor, M.S., Lam, W.L., Robinson, W.P., Brown, C.J.: Chromosome-wide dna methylation analysis predicts human tissue-specific x inactivation. Hum. Genet. **130**(2), 187–201 (2011)
8. Saito, Y., Kanai, Y., Sakamoto, M., Saito, H., Ishii, H., Hirohashi, S.: Expression of mrna for dna methyltransferases and methyl-cpg-binding proteins and dna methylation status on cpg islands and pericentromeric satellite regions during human hepatocarcinogenesis. Hepatology **33**(3), 561–568 (2001)
9. Esteller, M.: Cpg island hypermethylation and tumor suppressor genes: a booming present, a brighter future. Oncogene **21**(35), 5427–5440 (2002)
10. Gardiner-Garden, M., Frommer, M.: Cpg islands in vertebrate genomes. J. Mol. Biol. **196**(2), 261–282 (1987)
11. Hackenberg, M., Previti, C., Luque-Escamilla, P.L., Carpena, P., Martínez-Aroza, J., Oliver, J.L.: Cpgcluster: a distance-based algorithm for cpg-island detection. BMC Bioinformatics **7**, 446 (2006)
12. Chuang, L.Y., Huang, H.C., Lin, M.C., Yang, C.H.: Particle swarm optimization with reinforcement learning for the prediction of cpg islands in the human genome. PLoS ONE **6**(6), e21036 (2011)
13. Chuang, L.Y., Chang, H.W., Lin, M.C., Yang, C.H.: Chaotic particle swarm optimization for detecting snp-snp interactions for cxcl12-related genes in breast cancer prevention. Euro. J. Cancer Prev.: Official J. Euro. Cancer Prev. Organ. (ECP) **21**(4), 336–342 (2012)
14. González-Miranda, J.M.: Synchronization and Control of Chaos: An Introduction for Scientists and Engineers. Imperial College Press, UK (2004)
15. Chiang, Y.C., Chuang, L.Y., Yang, C.H., Lin, Y.D.: A hybrid approach for cpg island detection in the human genome
16. Peitgen, H.O., Jürgens, H., Saupe, D.: Chaos and Fractals: New Frontiers of Science. Springer-Verlag, New York (2004)

COSYS: A Computational Infrastructure for Systems Biology

Fabio Cumbo[1,2,5] (iD), Marco S. Nobile[3,5], Chiara Damiani[3,5],
Riccardo Colombo[3,5], Giancarlo Mauri[3,5], and Paolo Cazzaniga[4,5(✉)] (iD)

[1] Institute for Systems Analysis and Computer Science,
Italian National Research Council, Via Dei Taurini 19, 00185 Roma, Italy
fabio.cumbo@iasi.cnr.it
[2] Department of Engineering, Third University of Rome,
Via Della Vasca Navale 79, 00154 Rome, Italy
[3] Department of Informatics, Systems and Communication,
University of Milano-Bicocca, Viale Sarca 336, 20125 Milano, Italy
{nobile,chiara.damiani,riccardo.colombo}@disco.unimib.it
[4] Department of Human and Social Sciences, University of Bergamo,
Piazzale S. Agostino 2, 24129 Bergamo, Italy
paolo.cazzaniga@unibg.it
[5] SYSBIO.IT Centre of Systems Biology, Milano, Italy

Abstract. Computational models are essential in order to integrate and extract knowledge from the large amount of -*omics* data that are increasingly being collected thanks to high-throughput technologies. Unfortunately, the definition of an appropriate mathematical model is typically inaccessible to scientists with a poor computational background, whereas expert users often lack the proficiency required for biologically grounded models. Although many efforts have been put in software packages intended to bridge the gap between the two communities, once a model is defined, the problem of simulating and analyzing it within a reasonable time still persists. We here present COSYS, a web-based infrastructure for Systems Biology that guides the user through the definition, simulation and analysis of reaction-based models, including the deterministic and stochastic description of the temporal dynamics, and the Flux Balance Analysis. In the case of computationally demanding analyses, COSYS can exploit GPU-accelerated algorithms to speed up the computation, thereby making critical tasks, as for instance an exhaustive scan of parameter values, attainable to a large audience.

Keywords: Systems biology · Modeling and simulation · Flux Balance Analysis · GPGPU computing · High-performance computing

1 Scientific Background

Recent improvements in experimental -*omics* techniques applied to the study of biomolecular systems, generated an unprecedented amount of data. Disappointing naïve expectations, data alone are not able to account for the complexity of

© Springer International Publishing AG 2017
A. Bracciali et al. (Eds.): CIBB 2016, LNBI 10477, pp. 82–92, 2017.
DOI: 10.1007/978-3-319-67834-4_7

biological systems, pointing out that only quantitative methods are able to shed light on design principles and regulatory mechanisms behind these systems. In this context, Systems Biology took advantage of different competencies in order to develop quantitative and predictive methods based on computer-aided mathematical modeling.

In particular, the definition of an appropriate modeling framework is of paramount relevance to determine the usefulness of a model. Different mathematical formalisms have been proposed, which mainly differ in the level of detail and in the biological data required to build and parameterize the model, ranging from the coarse-grained (interaction-based, constraint-based) to the fine-grained (mechanism-based) approach, where mathematical models vary with respect to the size of the biological systems under investigation, the computational costs required to perform the analyses, and the nature of the computational results (i.e., qualitative or quantitative) [3].

The mechanism-based modeling approach has the greatest predictive capability concerning the functioning of the system at molecular level, as it can be exploited to perform time-course simulations, but has limited applicability in the context of genome-wide networks. On the other hand, when dealing with large metabolic networks, constraint-based models can facilitate the identification of pivotal components or modules of the system under investigation, although they neglect most of the quantitative and kinetic properties of its components and interactions. Indeed, constraint-based modeling mainly requires knowledge of the stoichiometry of the metabolic network to investigate functional steady states that a system is able to reach when subject to a set of constraints.

It is clear that to promote the widest use of these modeling approaches, they must be conveyed in the form of software packages, possibly using a standard notation. Although almost 300 different software compatible with the Systems Biology Markup Language (SBML) [6] have been currently released, only few exploit the advantages of a web interface: providing a non exhaustive list, it is possible to cite Biomodels, Cell Cycle DB, FAME, JWS Online, PathCase-SB (exploiting the RoadRunner simulation engine), SYCAMORE and The Virtual Cell (due to space constraints, we refer the reader to the list and links provided in http://sbml.org/SBML_Software_Guide/SBML_Software_Summary for details and comparison purposes).

Strikingly, none of them allows the execution of both mechanism-based and constraint-based simulations and most importantly (with the exception of Condor COPASI [7]) none of them are able to exploit high performance computing capabilities. For these reasons we developed COSYS, a web-based infrastructure for Systems Biology able to perform both constraint-based (FBA) and mechanism-based simulations (deterministic and stochastic), and to exploit the extensive computational power of Graphics Processing Units (GPU) to run the massive number of simulations required by computational tasks like the parameter scan. Moreover, acknowledging the pivotal importance of community efforts to model and simulate complex biological systems, COSYS extends the collaborative approach – proposed by Cell Collective [5] for models based on a qualitative mathematical formalism – to reaction-based quantitative models and pushes it forward by enabling the sharing of research projects either in a restricted workgroup or amid all COSYS users.

2   Materials and Methods

A non-spatial mechanism-based model (MBM) of a biochemical system can be formalized by specifying the set $\mathcal{S} = \{S_1, \ldots, S_N\}$ of molecular species occurring in the system, and the set $\mathcal{R} = \{R_1, \ldots, R_M\}$ of chemical reactions taking place among these species [4]. Reactions can be defined as

$$R_j : \sum_{i=1}^{N} \alpha_{ji} S_i \xrightarrow{k_j} \sum_{i=1}^{N} \beta_{ji} S_i,$$

where $\alpha_{ji}, \beta_{ji} \in \mathbb{N}$ are stoichiometric coefficients associated, respectively, with the i-th reactant and the i-th product of the j-th reaction, with $i = 1, \ldots, N$, $j = 1, \ldots, M$. According to the mass action kinetic assumption, the rate of a chemical reaction is proportional to the product of the concentrations of the reacting chemical species and $k_j \in \mathbb{R}^+$, the rate constant associated with reaction R_j. If we let $[S_i]$ be the concentration of reactant S_i, then reaction R_j occurs at a rate (or flux) $v_j = k_j \prod_{i=1}^{N} S_i^{\alpha_{ji}}$.

2.1   Biochemical Modeling and Simulation

Under the assumption that the biochemical system has constant temperature and volume, and all reactions follow the mass-action kinetics [4], any MBM can be converted into a set of coupled Ordinary Differential Equations (ODEs), which can then be simulated by means of an ODE numerical solver [1]. Among the existing integration algorithms (e.g., Euler's method, the family of Runge-Kutta methods), LSODA [14] is one of the most popular, thanks to its capability of dealing with stiffness by automatically switching between explicit and implicit integration methodologies.

When some chemical species are present in a few molecules, however, ODEs can fail to capture the emerging effects of biochemical stochastic processes [17] (e.g., multi-stability, state switching). In such a case, methods like the Gillespie's Stochastic Simulation Algorithms (SSA) [4] can be employed to accurately simulate the dynamics of the MBM.

SSA performs the simulation by calculating, at each step, the reaction to be fired and the time interval before the reaction actually occurs. Thus, SSA is a markovian process that can be computationally cumbersome for systems in which the biological noise is relevant and some of the reactions are characterized by a high firing rate. In this condition, however, the number of firings in a fixed time interval can be approximated by means of Poisson processes: this is rationale of methods like tau-leaping [2], which can be faster than standard SSA, under some circumstances.

The COSYS platform provides both deterministic and stochastic simulation, based on LSODA [14], SSA [4] and tau-leaping [2] algorithms. Moreover, COSYS offers the possibility of accelerating typical analysis methods—e.g., Parameter Sweep Analysis (PSA)—by leveraging coarse-grained GPU-accelerated simulators, which strongly reduce the overall running time of the analyses. Specifically, COSYS integrates the GPU-based deterministic simulator cupSODA [11] and the stochastic simulator cuTauLeaping [12].

2.2 Flux Balance Analysis

If a steady state is assumed for the amount $\#$ of each species in a MBM, i.e., $d\#S_i/dt = 0 \; \forall i$, then Flux Balance Analysis (FBA) can be applied to determine the flux distribution $v = (v_1, \ldots, v_M)$ that maximizes or minimizes the objective $Z = \sum_{j=1}^{M} w_j v_j$, where w_j is a coefficients that represents the contribution of flux j in vector v to the objective function Z.

Given a $N \times M$ matrix A, referred to as stoichiometric matrix, whose element a_{ji} takes value $-\alpha_{ji}$ if the species S_i is a reactant of reaction R_j, $+\beta_{ji}$ if the species S_i is a product of reaction R_j and 0 otherwise; the problem is postulated as a general Linear Programming formulation:

$$\text{maximize or minimize } Z$$
$$\text{subject to } Sv = 0, \; v_{min} \leq v \leq v_{max} \tag{1}$$

where v_{min} and v_{max} are two vectors specifying, respectively, the lower and upper boundaries of the admitted interval of each flux v_j. A negative lower bound indicates that flux is allowed in the backward reaction. The exchange of matter with the environment is represented as a set of unbalanced reactions (exchange reactions), enabling a predefined set of species to be inserted in or removed from the system. Boundaries on fluxes allow to reduce the degrees of freedom of the optimization problem, when the value of some variables is derived from experimental data, and/or incorporate other biological constraints, such as thermodynamics constraints on reaction reversibility, constraints on cell medium composition or intake and secretion rates, as well as enzymatic capacity constraints. For a more comprehensive description of FBA, the reader is referred to [13].

COSYS is able to execute the FBA of a MBM—as defined in Eq. 1—by exploiting the GLPK (GNU Linear Programming Kit) package.

3   Results

COSYS is a web-based platform for data intensive research concerning the field of Systems Biology. Its interface was conceived and developed following the current flat design principles to make the user interface as usable and intuitive as possible. Figure 1 shows the screenshots related to the main features offered by the COSYS user interface: definition of a model (top panel), simulation of the dynamics (middle panel), visualization of the results (bottom panel).

Fig. 1. The main features offered by the COSYS user interface: definition of a model (top panel), simulation of the dynamics (middle panel), visualization of the results (bottom panel). The COSYS user interface is characterized by a grid-based layout, and a clear minimalistic design to provide a good user experience and allow the users to focus on the features provided by the system.

Furthermore, as shown in Fig. 2, the COSYS infrastructure consists of three main technological components: (i) the storage unit; (ii) the relational database management system to keep track of users information, models and all performed simulations and analyses; (iii) the computational power unit based both on a CPU and a GPU, to perform single simulations or FBA analyses, and parallel tasks (such as parameter scans), respectively.

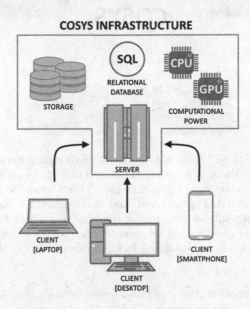

Fig. 2. COSYS infrastructure including the three main technological components: (i) the storage unit, (ii) the relational database management system to keep track of users information, models and the results of all performed simulations and analyses, and (iii) the computational power unit relying on a CPU and a GPU. Being a web-based platform, COSYS allows multiple accesses from a wide range of devices, from desktop and laptop computers to mobile devices.

COSYS relies on three major classes allowing the management, simulation and analysis of MBMs of biological systems. In particular, by following the typical Systems Biology approach, COSYS can be used to (i) create and modify MBMs, (ii) simulate the temporal dynamics by means of stochastic and deterministic algorithms, (iii) perform flux balance analyses, (iv) execute parameter sweep analyses, (v) visualize the simulation outcomes as well as other useful information (see Fig. 3).

MBMs can be defined and edited by means of an integrated user-friendly interface, which allows to specify the list of reactions involved in the biological process under investigation. For each reaction, the forward kinetic constant is required for the simulation of the dynamics of the model, while the backward kinetics constants can be optionally inserted in the case of reversible reactions.

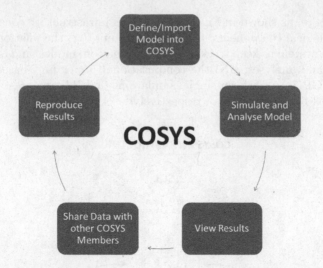

Fig. 3. COSYS working cycle that is typically followed by a generic user. COSYS is a "model-centric" infrastructure (i.e. every integrated tool works with a MBM); the user starts by uploading or creating a MBM with the COSYS (user-friendly) interface. The model can then be simulated or analyzed, and the obtained results can be visualized with an integrated fully interactive visualization tool. Both MBM and results can be then shared with other specific COSYS users or made available for the entire COSYS community, thus providing the possibility to reproduce all computational experiments.

Similarly, for each reaction, the lower and upper bounds, and the objective coefficient are required for the execution of FBA. Once a reaction is inserted into the model, COSYS automatically recognizes the molecular species and adds them into a specific table. It is clear that, for each molecular species, the initial concentration/amount is required to perform a simulation, either deterministic or stochastic. In addition, it is under development a functionality that will allow to import and export models written in SBML language, which will be automatically converted into the COSYS inner format. Note that, for what concern tasks that rely on the simulation of the dynamics, only SBML files of models that make use of the mass-action kinetics will be allowed.

Concerning the simulations and FBA tools integrated in COSYS, some functioning settings are required, i.e., the simulation time and the sampling interval in the case of the simulation of the dynamics; the choice whether to minimize or maximize the objective function in the case of FBA. Currently, COSYS allows the deterministic simulations by means of CPU-based implementations of LSODA algorithm [14] and stochastic simulations exploiting the Gillespie's stochastic simulation algorithm [4]; moreover, COSYS relies on GPU-based implementations of deterministic and stochastic algorithms: cupSODA [11] and cuTauLeaping [12], respectively, which are used to run intensive computational analyses requested by the users.

COSYS allows the execution of PSAs, both in the case of simulation and FBA tasks. In this context, besides the aforementioned functioning settings, the user must select the parameters to be perturbed: namely, kinetic constants and/or initial molecular concentrations/amounts in the case of simulation of the dynamics, and flux boundaries in the case of FBA. To be more precise, a PSA consists in the execution of a number of different simulations or FBA— based on the number of parameters that have to be perturbed and on the range variation—each one with a different value of the perturbed parameters. There- fore, a PSA can result computationally expensive, in particular in the case of complex MBMs characterized by high numbers of reactions and chemical species. Therefore, COSYS offers the possibility of strongly reducing the running time by means of GPU-powered versions of the simulation algorithms, specifically designed for demanding computational tasks like PSA.

To the aim of assessing the computational performances of COSYS, we performed a comparison with COPASI (often considered a "standard" for mechanism-based simulations in Systems Biology, because of the high number of research projects exploiting it), by considering the simulation time required to perform a PSA of the simplified model of glycolysis illustrated in [8], consisting of 8 reactions among 7 molecular species. In particular, we executed a PSA by varying the kinetic constant of the first reaction for a total of 1000 simulations. COPASI was executed on a MacBook Pro equipped with a 2.7 GHz Intel Core i5 CPU and 8 GB 1867MHz DDR3 RAM, while COSYS was executed exploit- ing an NVIDIA GeForce GTX 480 with 480 cores, 700 MHz of graphic clock, 1.4 GHz of processor clock, 1.5 GB GDDR5 of standard memory, and 1.8 GHz of memory clock. The running time of this PSA performed with COPASI was 0.447019 s, while the running time in COSYS was 0.004171 s thanks to cupSODA, the GPU-based implementation of the LSODA, thus achieving a significant 107× speed-up.

Concerning the graphical visualization of the simulation and FBA results, COSYS provides a set of tools: (i) the line plot that is used to graphically show the dynamics of all molecular species over time, with the possibility of choosing which species to include in the plot; (ii) the surface plot and phase space diagram for the visualization of the PSA results, in the case of the simulation of the dynamics of a specific MBM. For each visualization style, the plot is completely interactive. It is indeed possible to zoom-in or zoom-out, and scroll the resulting chart through the time axes. Additional and more advanced plotting capabilities will be introduced in the next releases.

A further relevant feature of COSYS is cooperation. As a matter of fact, the platform allows to create projects that are conceptually similar to folders whose permission can be set to private, shared, or public, and can contain mod- els, simulations and analysis results. On the one hand, in the case of *private* projects, only the owner can access and modify the contents, having the possi- bility of organizing all contents related to one or more MBMs. On the other hand, *shared* projects are visible to a restricted number of COSYS users, chosen by the author of the project. All COSYS users can browse and visualize *public* projects,

Table 1. Comparison of the main features of COSYS and COPASI. ● feature currently available; ★ feature currently under development; ○ feature planned for a future release.

	Feature	COPASI	COSYS
General software features	Web-based		●
	Freeware	●	●
	Open source	●	
	GUI	●	●
Interoperability	SBML Import/Export	●	★
	SED-ML Import/Export	●	
	Save/Share in public repositories		●
	Save/Share in private repositories		●
Usability	Model definition from empty project	●	●
	Model editing	●	●
	Import/export in software format	●	
	Graphical visualization of model structure		○
	Documentation accessible from the tool		●
	Wizard	●	
	Life-scientist oriented GUI	●	●
Simulation and analysis tools	LSODA algorithm	●	●
	SSA	●	●
	Tau-leaping	●	●
	Hybrid simulation algorithms	●	○
	GPU-powered simulation		●
	FBA		●
	Elementary Flux Modes	●	
	Sensitivity Analysis	●	○
	PSA	●	●
	Parameter estimation	●	○
	Reverse engineering		○
Data outputting and visualization	2D plotting	●	●
	3D plotting	●	●
	Phase space plotting	●	●
	Data saving in tabular format	●	●

accessing their contents and importing all MBMs, simulation and analysis results into their personal workspace, having the chance of reproducing all computational experiments.

Finally, in Table 1 we provide a comparison between the main features of COSYS and COPASI (the leading computational tool for Systems Biology), focusing on some keys aspects such as: general software features, interoperability, usability, simulation and analysis tools, data outputting and visualization.

4   Conclusion

In this paper we presented COSYS, an infrastructure for computational Systems Biology. COSYS allows to define mathematical models of biological systems by means of a user-friendly web interface. The infrastructure will also allows to import, export and share models, relying on the SBML standard: a full support for this useful exchange format is one of our main goals. Some advanced SBML features—like assignment rules and events—will be taken into account in future releases. Moreover, no spatial information can be specified in a model defined by COSYS, since appropriate a CPU and GPU-powered spatial simulators are currently under development.

For what concerns the computational investigations that can be realized with COSYS, it is currently possible to perform both deterministic and stochastic simulations of the system dynamics and to graphically visualize in the browser the obtained results. In particular, COSYS offers GPU-powered tools to the aim of accelerating the execution of computationally demanding tasks such as Parameter Sweep Analysis. In addition, a tool to perform the Flux Balance Analysis of the models is present, though no graphical visualization of the results is currently available.

As a future extension of COSYS, we will develop efficient hybrid algorithms for the simulation of the dynamics of models characterized by multiple temporal and numerical scales [15]; moreover, LASSIE, a GPU-based fine-grain solution for the simulation of large-scale MBMs will be included in COSYS [16]. COSYS will be further improved to guide the user among the different simulation algorithms available through the GUI; COSYS will automatically suggest the most fitting strategy, according to the characteristics of the MBM under investigation and the computational analysis required. In addition, we plan to include a tool to visualize the structure of the model, as well as to show the fluxes obtained by means of FBA. We will add tools for the Parameter Estimation of the kinetic constants [9] or the initial molecular amounts of the models; the reverse engineering of MBMs, according to experimental time-series [10]; the Sensitivity Analysis to help the researchers to find insights about the functioning of the biological system under investigation. COSYS will also exploit high performance computing to explore the space of feasible solutions of the FBA problem, taking into account alternative optimal solutions, as well as sub-optimal ones.

COSYS is developed using a *continuous delivery* philosophy: new advanced features will be incrementally added, tested and deployed in production, dynamically adapting the infrastructure to any request and feedback provided by the users.

Finally, the COSYS infrastructure for Systems Biology is available at the following address: http://www.sysbio.it/cosys; a free registration is required to access the platform.

Acknowledgments. This work has been supported by SYSBIO Centre of Systems Biology, through the MIUR grant SysBioNet—Italian Roadmap for ESFRI Research Infrastructures.

References

1. Butcher, J.C.: Numerical Methods for Ordinary Differential Equations. John Wiley & Sons, USA (2003)
2. Cao, Y., Gillespie, D.T., Petzold, L.R.: Efficient step size selection for the tau-leaping simulation method. J. Chem. Phys. **124**(4), 044109 (2006)
3. Cazzaniga, P., Damiani, C., Besozzi, D., et al.: Computational strategies for a system-level understanding of metabolism. Metabolites **4**(4), 1034–1087 (2014)
4. Gillespie, D.T.: Exact stochastic simulation of coupled chemical reactions. J. Comput. Phys. **81**(25), 2340–2361 (1977)
5. Helikar, T., Kowal, B., McClenathan, S., et al.: The cell collective: toward an open and collaborative approach to systems biology. BMC Syst. Biol. **6**(1), 96 (2012)
6. Hucka, M., Finney, A., Sauro, H.M., et al.: The systems biology markup language (SBML): a medium for representation and exchange of biochemical network models. Bioinformatics **19**(4), 524–531 (2003)
7. Kent, E., Hoops, S., Mendes, P.: Condor-COPASI: high-throughput computing for biochemical networks. BMC Syst. Biol. **6**(1), 1 (2012)
8. Klipp, E., Liebermeister, W., Wierling, C., et al.: Systems Biology: A Textbook. Wiley, San Diego (2009)
9. Nobile, M.S., Besozzi, D., Cazzaniga, P., Mauri, G., Pescini, D.: A GPU-based multi-swarm PSO method for parameter estimation in stochastic biological systems exploiting discrete-time target series. In: Giacobini, M., Vanneschi, L., Bush, W.S. (eds.) EvoBIO 2012. LNCS, vol. 7246, pp. 74–85. Springer, Heidelberg (2012). doi:10.1007/978-3-642-29066-4_7
10. Nobile, M.S., Besozzi, D., Cazzaniga, P., et al.: Reverse engineering of kinetic reaction networks by means of cartesian genetic programming and particle swarm optimization. In: Proceedings of CEC 2013, vol. 1, p. 1594–1601 (2013)
11. Nobile, M.S., Besozzi, D., Cazzaniga, P., et al.: GPU-accelerated simulations of mass-action kinetics models with cupSODA. J. Supercomput. **69**(1), 17–24 (2014)
12. Nobile, M.S., Cazzaniga, P., Besozzi, D., et al.: Cutauleaping: a GPU-powered tau-leaping stochastic simulator for massive parallel analyses of biological systems. PLoS ONE **9**(3), e91963 (2014)
13. Orth, J., Thiele, I., Palsson, B.: What is flux balance analysis? Nat. Biotechnol. **28**(3), 245–248 (2010)
14. Petzold, L.R.: Automatic selection of methods for solving stiff and nonstiff systems of ordinary differential equations. SIAM J. Sci. Stat. Comp. **4**(1), 136–148 (1983)
15. Salis, H., Kaznessis, Y.N.: Accurate hybrid stochastic simulation of a system of coupled chemical or biochemical reactions. J. Chem. Phys. **122**(5), 054103 (2005)
16. Tangherloni, A., Nobile, M.S., Besozzi, D., et al.: LASSIE: a large-scale simulator of mass-action kinetics models accelerated on graphics processing units. BMC Bioinform. **18**(1), 246 (2017)
17. Wilkinson, D.: Stochastic modelling for quantitative description of heterogeneous biological systems. Nat. Rev. Genet. **10**(2), 122–133 (2009)

Statistical Texture-Based Mapping
of Cell Differentiation Under Microfluidic Flow

Veronica Biga[1]([✉]), Olívia M. Alves Coelho[2], Paul J. Gokhale[3],
James E. Mason[3], Eduardo M.A.M. Mendes[4], Peter W. Andrews[3],
and Daniel Coca[5]

[1] Faculty of Biology, Medicine and Health, The University of Manchester,
Oxford Road, Manchester, M13 9PL, UK
veronica.biga@manchester.ac.uk
[2] Instiuto Tecnológico de Aeronáutica, São José dos Campos, SP 12228-900, Brazil
oma.coelho@gmail.com
[3] Centre for Stem Cell Biology, The University of Sheffield,
Western Bank, Sheffield S10 2TN, UK
{p.gokhale,james.mason,p.w.andrews}@sheffield.ac.uk
[4] Universidade Federal de Minas Gerais,
Antonio Carlos 6627, Belo Horizonte, Brazil
emmendes@ufmg.br
[5] Automatic Control and Systems Engineering Department,
The University of Sheffield, Mappin Street, Sheffield S1 3JD, UK
d.coca@sheffield.ac.uk

Abstract. Timelapse microscopy enables long term monitoring of biological processes, however a major bottleneck in assesing experimental outcome is the need for an automated analysis framework to extract statistics and evaluate results. In this study, we use Gabor energy texture descriptors to generate a high dimensional feature space which is analysed with principal component analysis to provide unsupervised characterisation of texture differences between pairs of images. We apply this technique to differentiation of human embryonic carcinoma cells in the presence of all-trans retinoic acid (RA) and show that differentiation outcome can be predicted directly from texture information. A microfluidic environment is used to deliver pulses of RA stimulation over five days in culture. Results provide insight into the dynamics of cell response to differentiation signals over time.

Keywords: Principal component analysis · Texture features · Gabor energy · Fate mapping · Cell differentiation

1 Introduction

Embryonic stem cells have the ability to generate all cells in the adult body through differentiation, a process by which cells acquire highly specialised function and morphology. In the developing embryo, cell differentiation undergoes

---

V. Biga and O.M. Alves Coelho have contributed equally to this work.

© Springer International Publishing AG 2017
A. Bracciali et al. (Eds.): CIBB 2016, LNBI 10477, pp. 93–106, 2017.
DOI: 10.1007/978-3-319-67834-4_8

,a complex spatial patterning with very precise timings difficult to replicate in vitro. As a result, differentiation of cells in culture is inefficient and not fully understood. Recent studies indicate that cell response to external factors can be amplified by pulsing of signals as opposed to constant exposure [1,2]. Two main technologies can be used to enable control of input signal properties, fast solution switching in microfluidic environment [1] and photo-activatable systems [2]. Microfluidics have the advantage of delivering multiple inputs and controlling other culture parameters such as shear stress and gradients [3].

In this study, we investigated the differentiation of NTERA2 [5,6] human embryonic carcinoma cells under controlled microfluidic flow. When exposed to all-trans retinoic acid, NTERA2 are known to differentiate towards a non-neural fate identified by surface marker ME311 and neural cells detected after 3–4 weeks [7], however expression at early stages in differentiation is not well characterised. We monitored cell differentiation using a timelapse microscope which enables acquiring a timeseries of images at set time intervals thus capturing not only the final outcome but the entire process of cells starting from a proliferative state and leading into early differentiation in the presence of a morphogen over several days. Timelapse produces extensive video data which needs to be processed in an automated way in order to extract useful statistics and characterise the differentiation process over time.

The problem of automating statistical analysis of cell differentiation is extremely challenging primarily due to lack of statistical descriptors to describe changes occuring in culture over time. Recent studies have shown that texture descriptors can enable classification of human stem cells using phase contrast microscopy [4] however statistics over time are not discussed. From an experimental point of view some key questions regarding differentiation are: (i) when does onset happend; and (ii) did certain differentiation conditions favour a particular cell type being produced. In this study, we aim to address these questions in the context of early differentiation of human carcinoma cells by automated detection of changes in cell culture morphology over time based on texture information. Firstly, we describe a generic approach to qualitatively assess differences in textural characteristics of images from the Brodatz dataset for a fixed timepoint. We then extend this approach to quantitatively monitor changes in a timeseries of images through an unsupervised dimensionality reduction technique.

2   Principal Component Analysis-Based Texture Discrimination Technique

2.1   Gabor Energy Texture Features

Gabor filters [8] produce a decomposition of intensity values in an image $I(x, y)$ into sub-bands with preffered orientation and sparial frequency, revealing hidden spatial information by kernel convolution:

$$g_{\lambda,\sigma,\gamma,\theta,\varphi}(x,y) = e^{-\frac{x'^2+\gamma^2 y'^2}{2\sigma^2}} \cos\left(2\pi \frac{x'}{\lambda} + \varphi\right); r_{\lambda,\sigma,\gamma,\theta,\varphi} = I * g \qquad (1)$$

where $x\prime = (x - x_0)\cos\theta + (y - y_0)\sin\theta; y\prime = -(x - x_0)\sin\theta + (y - y_0)\cos\theta$ and (x_0, y_0) represent the kernel centerl; $\theta \in [0\ \pi)$, λ and $\varphi \in (-\pi\ \pi]$ represent angle, frequency and phase respectively. The envelope is characterised by ellipticity γ and size σ linked through bandwidth b: $\frac{\sigma}{\lambda} = \frac{1}{\pi}\sqrt{\frac{\ln 2}{2}\frac{2^b+1}{2^b-1}}$. A series of Gabor filters computed for different parameter combinations are shown in Fig. 1.

Texture information in the image can be characterised by Gabor energy features computed as:

$$e_{\lambda,\sigma,\gamma,\theta}(x,y) = \sqrt{r^2_{\lambda,\sigma,\gamma,\theta,0}(x,y) + r^2_{\lambda,\sigma,\gamma,\theta,-\frac{\pi}{2}}(x,y)} \ . \tag{2}$$

In the following, we refer to texture features as a set of Gabor energy texture features generated for each image using parameters:

$$\lambda = [1/15; 1/30; 1/60; 1/120; 1/240]$$
$$\theta = [0; \pi/4; \pi/2; 3\pi/4; \pi; 5\pi/4; 6\pi/4; 7\pi/4] \tag{3}$$
$$\sigma \in [0.002, 0.03]$$
$$\gamma \in [0.001; 0.01]$$
$$b = 1$$

and organised into a texture feature dataset:

$$\mathbf{f}(x,y) = [\mathbf{f}^1(x,y), \mathbf{f}^2(x,y)\ldots\mathbf{f}^N(x,y)]; \overline{\mathbf{f}^{1:n}} = e_{\lambda_k,\sigma_k,\gamma_l,\theta_m}(x,y)\forall k,l,m. \tag{4}$$

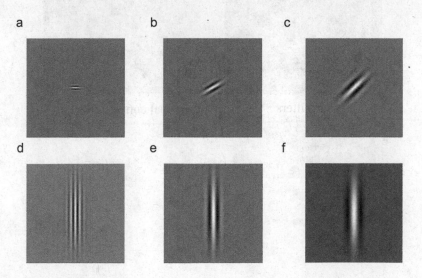

Fig. 1. Gabor filters generated for different orientation, size and bandwidth. Phase offset is set to $\varphi = 0$ throughout examples. (a) $\sigma = 5; \theta = 0; \gamma = 0.5; b = 1$; (b) $\sigma = 10; \theta = \pi/6; \gamma = 0.5; b = 1$; (c) $\sigma = 15; \theta = \pi/4; \gamma = 0.5; b = 1$; (d) $\sigma = 15; \theta = \pi/2; \gamma = 0.24; b = 0.5$; (e) $\sigma = 15; \theta = \pi/2; \gamma = 0.24; b = 1$; (f) $\sigma = 15; \theta = \pi/2; \gamma = 0.24; b = 2$.

2.2   PCA-Based Analysis Using Random Window Sampling

Extracting texture by Gabor energy features is a task of high computational complexity. To prevent this from becoming intractable for large images, texture was analysed in a number of 50 windows collected at random locations in the image. For each sampling window i and a given parameter combination j the Gabor energy output is integrated to produce texture observations: $Z(i,j) = \int \int \mathbf{f}^j(x,y)dxdy$. Thus for each image, a texture dataset of 50 observations by 280 features was extracted. Where multiple images and timepoints were considered, texture data was concatenated in a single large dataset and analysed with Principal Component Analysis (PCA) [9]. The Matlab implementation *princomp.m* was used to obtain the projection $W = A^T Z$ and data was visualised in the reduced principal component space.

2.3   Unsupervised Texture Discrimination of Brodatz Images

Images Can Be Distinguished by Texture in PCA-Reduced Space. Images from the Brodatz database [10] were analysed with the PCA-based technique using texture (Fig. 2). A combined texture feature set was produced by

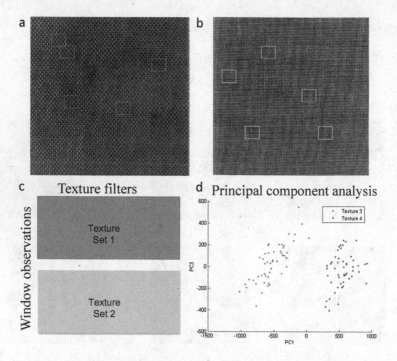

Fig. 2. PCA-based image analysis technique. (a) Sampling from Texture 1 at locations in red; (b) sampling from Texture 2 at locations in blue; (c) Combined feature dataset organised by window number vs filter number; (d) PCA analysis of all observations colour coded according to Texture Set 1 (red) compared to Texture Set 2 (blue). (Color figure online)

concatenating observations from the two texture for all filter parameters. The PCA analysis and 2D projections in the principal component space show that observations from the two images clearly segregate into separate groups (Fig. 2d). Thus without any prior information of the images PCA-based analysis of texture could be used to discriminate between two images.

One commonly used feature to describe an image is the intensity histogram which does not contain spatial information. However, distinct images can show similarity in grayscale intensity distributions. To highlight differences between a texture-based approach and grayscale intensity, we applied the Gabor energy and PCA-based analysis with random sampling to images with similar distributions represented as histograms (Fig. 3). Indeed even in this case, on the basis of texture information, PCA can discriminate between images with distinct texture characteristics without prior knowledge of the data (Fig. 3a). Only few observations are overlapping and a separation boundary can be found.

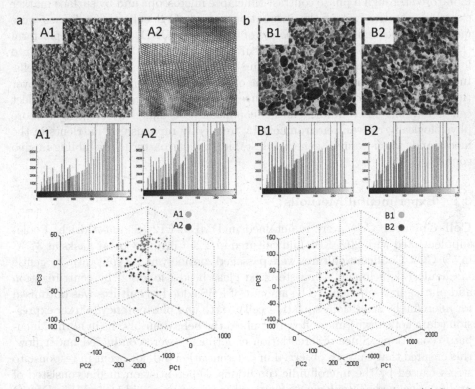

Fig. 3. PCA-based discrimination of textured images from Brodatz dataset. (a) Set of two different textured images A1 and A2 (top panel), grayscale intensity histograms (middle panel) and PCA visualisation show minimal overlap between samples from A1 and A2 (bottom panel); (b) Set of highly similar textured images B1 and B2 (top), grayscale intensity histograms (middle) and PCA analysis showing complete overlap between observations from B1 and B2 (bottom).

Texture Similarity Between Images. In assessing similarity it is critical to be able to determine when images are identical in texture. When tested on images acquired from the same textured source material, the PCA-based approach indicated complete overlap between texture observations (Fig. 3b).

Taken together, these results indicate that a PCA-based approach for analysing texture can describe both similarity and differences between images. In the following, we apply this technique to describe the timeline of differentiation in human embryonic carcinoma cells in the presence of a morphogen.

3   Differentiation of NTERA2 Embryonic Carcinoma Cells Induced by Retinoic Acid

Embryonic carcinoma cells from the NTERA2 cl.D1 [5,6] cell line were exposed to all-trans retinoic acid (RA) over 5 days in culture and differentiation was monitored through a phase contrast timelapse microscope and by surface marker expression at the end of the incubation period. Cell culture conditions were maintained over time through the use of the Cellasic ONIX microfluidic platform [13]. This system uses pressure to accurately control the delivery of media into four independent culture chambers and solution switching from multiple wells. In addition to controlling the amount of media and frequency of delivery over time, the system is optimised to produce minimal shear stress which could affect cell behaviour. Microfluidic culture conditions are aimed at creating a more physiologically relevant environment by frequently replenishing nutrients in the media and (when desired) generating stable gradient patterns resembling in vivo conditions.

3.1   Experimental Methods

Cell Culture. Cells were maintained in DMEM-F12 (Sigma-Aldrich, Poole) supplemented with 20% FBS and cultured in T25 Corning Costar flasks at 37 °C in 5% CO2. Confluent cultures were passaged routinely at 1:3 split ratio by gentle mechanical detachment using sterilised glass beads, followed by centrifugation and plating in fresh media. The amount of FBS added to cultures was optimised for microfluidic conditions such that cell growth and pluripotency marker expression (data not shown) under flow resembled the behaviour of cells in normal incubator conditions. Since differentiation experiments were conducted under flow, this ensured that changes observed in differentiation did not occur in response to stress caused by the microfluidic conditions. Differentiation media consisted of proliferative media supplemented with all-trans retinoic acid (10^{-2}M) in DMSO (Sigma-Aldrich) at a concentration of 10^{-7}M (RA-7).

Cell Differentiation. Cells were cultured in M04S (Merck Millipore) mammalian culture plates precoated with Matrigel (BD Biosciences). Coating involved shallow filling of culture chambers with cooled Matrigel solution by gravity flow or automated loading on ice. Coated plates were maintained at room temperature for minimum 1h prior to incubation. Single cells were dissociated using trypsin, resuspended at 2–3 million cells per ml and loaded automatically under high flow. Following a 2h attachment period, cultures were fed every 1h at 0.25psi flow rate. The delivery of media to the cells was controlled automatically to create 20 min pulses of RA-7 every 1h or 3h repeated over a 5 day incubation period. Continuous conditions indicate that RA-7 is always present in the media. Detailed protocols are included in Appendix.

Timelapse Microscopy. Cultures were monitored by phase-contrast microscopy using an IX70 (Olympus) inverted microscope fitted with a motorised XY stage and focus controller(Prior Scientific) controlled by a Proscan III automation controller (Prior Scientific). Cells were imaged using a 10x, 0.45 NA phase contrast object (Olympus) and detected using a Photometrics Evolve 512 EMCCD camera (Photometrics) every 10 min, controlled via Micromanager [11].

Immunocytochemistry. At the end of the five day incubation period, cells were fixed in 4%PFA, blocked with 5% FBS in PBS, incubated with primary monoclonal antibody for ME311 (9-O-acetyl-GD) and Hoechst33342 (Life Technologies) then incubated with Alexa Fluor 647 (Thermo-Fisher) conjugated secondary antibody. Protocol was performed in M04S microfluidic plates following the automated immunostaining surface marker protocol [12]. Plates were imaged using the InCell Analyzer2000 (GE Healthcare) and the number of ME311 positive cells was counted by image processing using Developer Toolbox.

3.2 Microfluidic Frequency Control of Retinoic Acid (RA-7) Ellicits Differential Response in Cell Fate

Cells maintained in media containing RA-7 exhibited varying levels of differentiation over five days in culture (Fig. 4). Although cells in all conditions proliferated and became confluent, the frequency of RA-7 greatly affected the proportion of cells expressing the surface marker ME311 (Table 1). Adding RA-7 at a 1:3 dilution every 1h (Fig. 4: 1h/3p) as a pulse caused only 9% of cells to express ME311 (Table 1: 1h/3 pulse). However, when the same overall concentration of RA-7 was added as a single pulse every 3h (Fig. 4: 3hp), the proportion of ME311+ cells increased to $27 \pm 14\%$ (Table 1: 3h pulse). This indicated that cell response to morphogen signal RA is not integrated over time and differentiation is enhanced at high concentrations.

We hypothesised that more frequent addition while keeping the concentration fixed would lead to increased differentiation potential. Indeed, when increasing the frequency of adding RA from every 3h to every 1h, we noted an 18% increase in ME311+ cells up to $48 \pm 4\%$ (Table 1: 3h pulse vs 1h pulse). These results showed nonlinear response, i.e. when frequency is increased 3x, the number of ME311+ cells increased 1.8x. Remarkably, in continuous RA-7, the proportion of cells expressing ME311 dropped by 18% compared to pulsing (Table 1: 1h cont vs 1h pulse).

Table 1. Proportion of cells expressing surface marker ME311 in different flow conditions

Condition	Nuclei counts	ME311+	%ME311+
1h/3 pulse	10122	910	9%
3h pulse	6269	1058	27 ± 14%
	8695	3217	
1h pulse	6913	3122	48 ± 4%
	9850	8046	
1h cont	8201	2460	30%

The continuous condition best resembles classical differentiation experiments where the morphogen is always present in the media. Results indicated that pulsing at particular frequencies increases differentiation potential. The mechanism by which this occurs is undetermined but could involve receptor saturation when the morphogen is always present while the receptor continues to respond in pulsing conditions. Overall, levels of ME311 showed that cell response appeared nonlinear with respect to frequency and concentration of morphogen. However, the timeline of differentiation remained unknown. In the following, the PCA-based analysis strategy is used to describe differentiation potential from video data acquired throughout the differentiation process.

4   PCA-Based Analysis of Cell Differentiation from Timeseries of Images

The most significant fate shift in surface marker expression was detected between the 1h pulsed and 3h pulsed conditions (Fig. 4 and Table 1). We used timelapse microscopy to further investigate differentiation onset of these cultures over the entire 5 day period. A total of 721 images were acquired from culture chambers exposed to RA-7 every 1h and every 3h. We acquired 5 positions from each chamber representing technical replicates. All images (from each position and timepoints) were sampled for texture at 50 random locations per image yielding an overall dataset of 500 texture observations by 280 features. Texture data was extracted using Gabor energy and analysed with PCA as described in Sect. 2.2. In addition, a linear discriminant analysis was applied.

4.1   Computational Methods

Cross-Validation of Linear Discriminant Classifier. A linear discriminant classification technique applied to the PCA-projected data was used to separate texture information acquired from different images and implemented in Matlab using *classify.m*. The classifier was optimised using n-fold cross validation: (i) data was divided into four sets; (ii) the linear discriminant was trained on one set

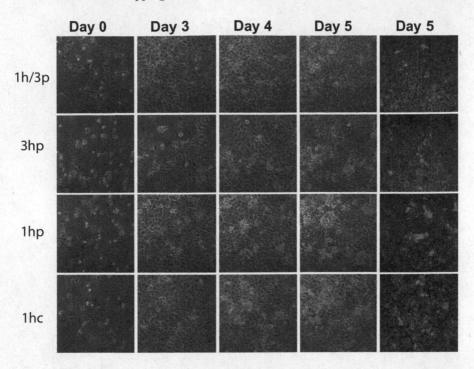

Fig. 4. Cell differentiation under microfluidic flow. Representative images from time-lapse experiments carried out at set frequencies of RA-7, every 1h pulse of RA-7/3 (1h/3p); every 3h pulse (3hp); every 1h pulse (1hp); every 1h continuous (1hc); Left panels show antibody staining for surface marker ME311 (red) and nuclear marker Hoechst33342 (blue). (Color figure online)

and tested on the remaining three at all possible combinations; (iii) the classifier with overall best rate was chosen.

Optimisation of Window Size. Window size affected the classification results and needed to be optimised. To account for this, the global classifier that produced >95% classification rates at the smallest % coverage (image area covered by windows/total image area x100) of an image were chosen. This guarantees that images are not over-sampled. Overall, high classification rates (>95%) could be obtained from 10% coverage of images highlighting inherent redundancy in texture information across the image.

4.2   Texture Characteristics Reveal the Onset of Differentiation

Texture information was analysed with PCA yielding a representation of cell culture differentiation at Days 1 to 5 as observations projected in the principal component space (Fig. 5a). The dominant principal components accounted for most variability in the data with approximately 91% and 93% for two and

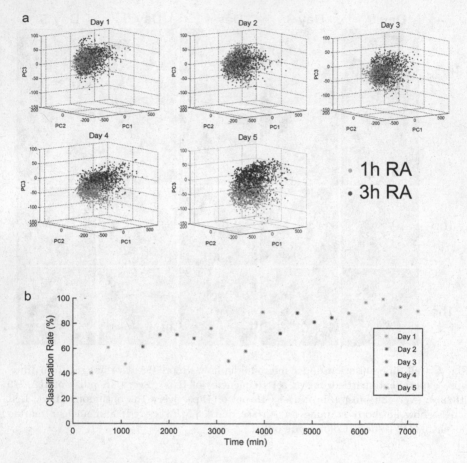

Fig. 5. Texture analysis of differentiation experiments. (a) Texture characteristics extracted from two culture conditions 1hRA (red) pulse and 3h RA pulse (blue) are visualised in the principal component space defined by the dominant three components; (b) Classification rates for linear discriminant trained on the last timepoint describe the divergence between the two clusters at 6h intervals. (Color figure online)

three components respectively; the remaining 7% of variability was split between remaining principal components. Thus we considered that the dominant three principal components are sufficient to describe the differences between the two conditions.

As expected, the cultures appeared unchanged in the first days post-plating. The first detectable differences in texture characteristics appeared in Day 3 and continued to amplify in Days 4 to 5. Specifically, compared to the location of clusters representing 1h RA and 3h RA at Day 1, at later incubation times the two conditions appeared to move in opposite directions in terms of texture characteristics leading to best separation at Day 5. The gradual shift observed in Days 3 to 5 indicate that after the onset of differentiation at approx 72h post-plating, the cells in the two flow conditions continue to diverge over time. It remains unclear whether the time period of 72h is related to cell cycle time (approximately 18h) in which case cells would require 3 to 4 divisions before committing to differentiation.

4.3   Timeline of Differentiation Is Mapped by Statistical Analysis

We performed a statistical analysis of texture characteristics over time through linear discriminant and cross-validation analysis. Because differentiation was validated through an independent antibody staining technique at the end of the incubation period (Day 5), we used pairs of images from the 1h and 3h conditions at the last timepoint to train and test a linear discriminant classifier with the cross-validation techniques described in Sect. 4.1.

The unsupervised classification of texture information is computed at each timepoint between the 1h and 3h RA conditions thus generating a quantitative measure of differentiation over time (Fig. 5b). As expected, at the initial timepoints, classification rates were approximately 50% indicative of high similarity in texture characteristics between the two conditions. This results is consistent with no differentiation.

At Day 3 classification rates started to exceed 75% but remained variable indicating that in some images, it is possible to discriminate between the two conditions while in others changes due to differentiation are unclear. We concluded that Day 3 marks the onset of differentiation. Furthermore, starting from 72h post-plating, the classification rate was persistently above 75% indicating that in Days 4 and 5, the two conditions produce clear differences in texture and this is consistent with surface marker expression levels at the end. Thus in Days 4 and 5 changes in texture induced by differentiation are permanent and continue to develop amplifying differences between the two cultures.

5   Discussion

In this study, we described a random sampling technique for extracting texture information from imaging data using Gabor energy and analysing differences using conventional PCA. In separate examples, we showed that this framework can successfully discriminate between distinct Brodatz textures but also detect similarity. By combining PCA analysis of texture with a linear discriminant, an unsupervised technique for measuring textural differences in a set of images is constructed. We apply this technique to a large imaging dataset showing differentiation of human embryonic carcinoma cells.

Using a microfluidic perfusion environment, we designed controlled variations in media conditions over five days. Differentiation potential was monitored only at the end of the experiment using the non-neural surface marker ME311. Experimental results showed that frequency patterns led to a significant change in the amount of differentiated cells expressing ME311 most evident between 1h and 3h pulsed conditions. Cell response was highly nonlinear with concentration and frequency. Remarkably, the behaviour of cells exposed continuously to RA diverged greatly from cells exposed to pulses suggesting that saturation may occur when the trigger molecule is always present in the media.

Principal components provided a representation of cell differentiation as linear combinations of texture filters applied to images. Gabor texture filters capture spatial information of changes in morphology that the cultures experience however these cannot be directly related to a biological mechanism. Nevertheless, pattern recognition techniques could be used to closer investigate a potential mapping between texture and morphology. Based on timelapse observations, morphology changes occured primarily from local changes in cell density. In the presence of retinoic acid, some of the cells in the same cultures became unproliferative forming valleys of darker regions while others continued to proliferate in brighter ridges surrounding unproliferative cells. These effects gave rise to irregular morphology when compared to undifferentiated cultures. Therefore, links between the texture characteristics and cell morphology could be investigated through the use of a live nuclear stain and markers associated with cell proliferation.

To investigate the timeline of differentiation, we analysed texture differences in images collected over the entire 5 day differentiation experiment. The results showed that a shift in differentiation outcome could be predicted directly from texture information. The trajectory of clusters containing observations from the two conditions was indicative of a fate shift away from the appearance at 48h post-plating (undifferentiated) and also diverging from each other. Since ME311 was favoured in the 1h RA condition, a potential mechanism by which cells in the 3hRA undergo a fate shift is by expressing higher levels of the neuronal markers. The trends in classification rates over time, provide a quantitative description of the timeline of differentiation which is unsupervised and label free. However, more advanced kernel-based PCA techniques and wavelet methods may provide additional insight. The statistical and experimental techniques described in our study provides the basis for automated analysis and monitoring of cell cultures in a variety of conditions that affect culture morphology and cell fate.

Acknowledgments. This work was funded by a Human Frontier Science Program grant. OC was funded by Conselho Nacional de Desenvolvimento Científico e Tecnológico, Brazil.

Appendix

Microfluidic experiments on the M04S mammalian plate were prepared according to the diagram in Fig. 6 and in addition the waste wells 7 were filled with 300 μl media per well to prevent gravity flow. The protocol for the microfluidic system is summarised in Algorithm 1.

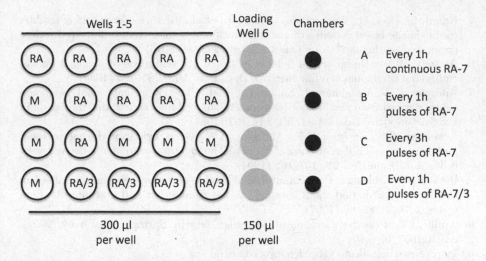

Fig. 6. Preparation of differentiation experiments in microfluidic M04S plate. Legends denote: (M) DMEM-F12+20% FBS media; (RA) all-trans retinoic acid at 10^{-7} in media M; (RA/3) RA diluted at 1:3 in media M.

Algorithm 1. Cell differentiation protocol for Cellasic ONIX

settemp 37.5
setflow X 1
setflow Y 1
% attachment stage
wait 240
% run for Vx as V2, V3, V4, V5 sequentially
open Vx
wait 5
close Vx
wait 15
open V1
wait 0.2
close V1
wait 39.8
% repeat sequences V2V1V3V1V4V1V5V1 for 120h in total

References

1. Sorre, B., Warmflash, A., Brivanlou, A.H., Siggia, E.D.: Encoding of temporal signals by the TGF-β pathways and implications for embryonic patterning. Dev. Cell **30**, 334–342 (2014)
2. Sokolik, C., Liu, Y., Bauer, D., McPherson, J., et al.: Transcription factor competition allows embryonic stem cells to distinguish authentic signals from noise. Cell Syst. **1**, 117–129 (2015)
3. Mehling, M., Savas, T.: Microfluidic cell culture. Curr. Opin. Biotechnol. **25**, 95–102 (2014)

4. Nanni, L., Paci, M., Caetano dos Santos, J., et al.: Texture descriptors ensembles enable image-based classification of maturation of human stem cell-derived retinal pigmented epithelium. PLoS One **11**(2), e01493399 (2016)
5. Andrews, P.W.: Retinoic acid induces neuronal differentiation of a cloned human embryonal carcinoma cell line in vitro. Dev. Biol. **103**, 285–293 (1984)
6. Andrews, P.W., Damjanov, I., Simon, D., et al.: Pluripotent embryonal carcinoma clones derived from the human teratocarcinoma cell line Tera-2: differentiation in vivo and in vitro. Lab. Invest. **50**, 147–162 (1984)
7. Ackerman, S.L., Knowles, B.B., Andrews, P.W.: Gene regulation during neuronal and non-neuronal differentiation of NTERA2 human teratocarcinoma-derived stem cells. Mol. Brain Res. **25**, 157–162 (1994)
8. Petkov, N., Kruizinga, P.: Computational models of visual neurons specialised in the detection of periodic and aperiodic visual stimuli: bar and grating cells. Biol. Cybern. **76**(2), 83–96 (1997)
9. Jolliffe, I.T.: Principal Component Analysis, 2nd edn. Springer, New York (2002). doi:10.1007/b98835
10. http://www.ux.uis.no/~tranden/brodatz.html
11. Edelstein, A.D., Tsuchida, M.A., Amodaj, N., et al.: Advanced methods of microscope control using μManager software. J. Biol. Methods **1**(2), e10 (2014)
12. https://www.merckmillipore.com
13. http://www.biocompare.com/Application-Notes/126441-Microfluidic-Perfusion-Enables-Long-Term-Cell-Culture-Precise-Microenvironment-Control-And-Gene-Expression-Analysis/

Constraining Mechanism Based Simulations to Identify Ensembles of Parametrizations to Characterize Metabolic Features

Riccardo Colombo[1,2]($\boxtimes$) (iD), Chiara Damiani[1,2], Giancarlo Mauri[1,2], and Dario Pescini[1,3]

[1] SYSBIO.IT Centre of Systems Biology,
Piazza Della Scienza 2, 20126 Milano, Italy
[2] Department of Informatics, Systems and Communication,
University of Milano-Bicocca, Viale Sarca 336, 20125 Milano, Italy
riccardo.colombo@disco.unimib.it,
{chiara.damiani,giancarlo.mauri}@unimib.it
[3] Department of Statistics and Quantitative Methods, University of Milano-Bicocca,
Via Bicocca Degli Arcimboldi 8, 20126 Milano, Italy
dario.pescini@unimib.it

Abstract. Constraint-based approaches have been proven useful to determine steady state fluxes in metabolic models, however they are not able to determine metabolite concentrations and they imply the assumption that a biological process is optimized towards a given function. In this work we define a computational strategy exploiting mechanism based simulations as a framework to determine, through a filtering procedure, ensembles of kinetic constants and steady state metabolic concentrations that are in agreement with one or more metabolic phenotypes, avoiding at the same time the need of assuming an optimization mechanism. To test our procedure we exploited a model of yeast metabolism and we filtered trajectories accordingly to a loose definition of the Crabtree phenotype.

Keywords: Systems biology · Mechanistic simulations · Steady state · ODEs · Fluxes · Ensembles · Kinetic parameters · Metabolism

1 Scientific Background

It is nowadays evident that biological processes must be described in terms of complex networks of non linear interactions involving several entities (genes, transcripts, proteins, metabolites) giving rise to emergent behaviors. This awareness, coupled with the fact that biological complex systems can be effectively analyzed only by means of mathematical modeling and simulations, gave rise during the last two decades to Systems Biology, a new discipline integrating computational modeling and "wet" experimental approaches [3]. In particular the study of metabolism has widely took advantage of Systems Biology approaches

© Springer International Publishing AG 2017
A. Bracciali et al. (Eds.): CIBB 2016, LNBI 10477, pp. 107–117, 2017.
DOI: 10.1007/978-3-319-67834-4_9

usually describing metabolic networks as hypergraphs in which nodes represent metabolites and edges indicate reactions [22].

Simultaneously to the development of computational techniques, progresses in high throughput technologies opened the "omics data" era characterized by a thrive of genome-scale metabolic reconstructions tailored on different cell types (unicellular organisms [2], healthy and diseased tissues [9]). However, technological limitations do not allow yet to investigate these genome-scale models by means of mechanism-based approaches (i.e., by simulating their temporal dynamics) [5]. For this reason, these models are currently studied through constraint-based approaches [11] exploiting information on metabolic network structure and assuming a pseudo-steady state for internal metabolites, thereby disregarding temporal evolution of the system: focusing on the metabolic steady state has been proven to be a valid assumption due to experimental evidences showing that *in vivo* metabolism reaches the steady state in few seconds [21].

The stoichiometric information retrievable from the structure of the metabolic network is the core of constraint-based modeling, indeed the stoichiometric matrix associated to a metabolic network mathematically defines changes in metabolites quantities when reactions are applied. The imposition of mass balance and of additional constraints (e.g., irreversibilities or boundaries on fluxes) allows to determine a feasible solution space containing flux distributions (i.e., flux values for each reaction in the model) reachable by the system and representing different functional states. Lastly, under the assumption that the metabolic system is optimized towards a given goal, optimization methods as flux balance analysis (FBA) [17], can be used to determine an optimal flux distribution that maximizes or minimizes a given metabolic task determined by an objective function (OF).

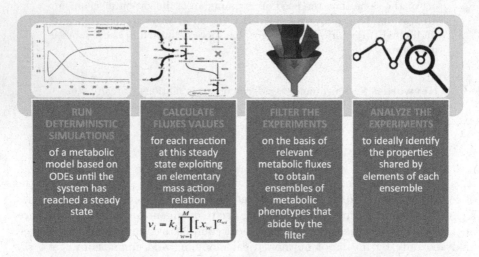

Fig. 1. Schematic workflow illustrating the four main phases of the computational procedure.

Recent studies [10] showed how the selection of the appropriate OF is essential when performing FBA investigations. This is due to the fact that often it is not possible to determine the formulation of the OF, but also to the fact that it is not possible to determine if the system is found in a sub-optimal state.

In a previous work, using an extension of FBA that we named Ensemble Evolutionary FBA (eeFBA) [6], we analyzed the capability of a cell to pursue alternative metabolic behaviors by altering its fluxes, or in other words, we determined which flux distributions are able to give rise to a specific metabolic behavior. To perform the analyses we generated a set of random OFs to be optimized by means of linear programming. We then filtered them accordingly to the definition of different metabolic phenotypes in order to obtain distinct ensembles of solutions that comply with defined phenotypes.

However, both with FBA and eeFBA it is not possible to determine the extent of metabolic concentrations when the system is at steady state due to the lack of information on kinetic constants. In the present paper we propose a novel strategy, where the ensembles of alternative phenotypes are still populated according to fluxes properties but extracting steady states from mechanism based simulations parametrized using initial concentrations retrieved from the literature and randomly sampled kinetic constants.

By doing this, we are able to infer ensembles of metabolite concentrations at steady state that are in accordance with a given metabolic phenotype independently from the definition of an appropriate OF and from the assumption that the cell is optimizing towards a specific objective.

2   Materials and Methods

With the procedure here devised and illustrated in Fig. 1, we setup several "experiments" that we define as follows: for each random parametrization we execute a number $(\mathcal{N})$ of simulations using, for each of them, a different but constant concentration of nutrient (e.g., glucose).

To perform the procedure, we firstly run deterministic simulations of a metabolic model based on ODEs and exploiting the LSODA solver until the system has reached a steady state. We then calculate fluxes values v_i for each reaction i at this steady state exploiting an elementary mass action relation:

$$v_i = k_i \prod_{w=1}^{M} [\chi_w]^{\alpha_{wi}} \tag{1}$$

where k_i is the rate constant of reaction i, $[\chi_w]$ is the concentration of species w and α_{wi} the stoichiometric coefficient with which species w participate to reaction i. At this stage, we filter the experiments on the basis of relevant metabolic fluxes to obtain ensembles of metabolic phenotypes that abide by the filter. In particular, in this work, to filter the experiments we used the same phenotype definition already published in [6]. The last step of the procedure is to analyze the experiments to ideally identify the properties shared by elements of each ensemble such as the presence of putative subphenotypes.

To test the procedure herein developed we defined the metabolic phenotype expressed by the Crabtree effect [8], a well known biological phenomenon taking place in some yeasts like *Saccharomyces cerevisiae*, implying a production of ethanol by fermentation when high concentrations of glucose are available in the extracellular environment preferring fermentation—regardless the availability of oxygen—with respect to the more energetically efficient oxidative phosphorilation (OXPHOS). In our test case we call "Crabtree-positive" the phenotype exhibiting the enhanced fermentation. On the same line, we call "Crabtree-negative" the phenotype of those yeasts (like *Kluyveromyces*) not showing the peculiar experimental characteristics.

To evaluate the effectiveness of the procedure in discriminating the two phenotypes and in selecting corresponding ensembles of kinetic constants and steady state metabolic concentrations, we used a simplified model of yeast metabolism (illustrated in Fig. 2) developed in [6] and designed to take into account only those pathways (metabolites and reactions) involved in the emergence of the Crabtree effect (CE).

To determine the initial concentrations of metabolites involved in the yeast model, we mined the literature and we set them accordingly to the average values illustrated in Smallbone et al. [20], and Canelas et al. [4]. From the *in vivo* experimental point of view this effect can be observed as the concomitant presence of alcoholic aerobic fermentation and reduction of OXPHOS rate when the glucose uptake from the medium progressively increases (e.g. by means of incremental glucose pulses added to yeast medium culture). The given biological definition of the CE however must be mathematically translated in order to formally and unequivocally determine metabolic response constraints defining the Crabtree-positive (C⊕) and Crabtree-negative (C⊖) phenotypes.

To this end we evaluated fluxes that in the model are proxies for OXPHOS— the sum of fluxes (v_o) for the two reactions summing up respiration, illustrated with A and B in Fig. 2— and alcoholic fermentation—ethanol secretion flux (v_e) represented with C in Fig. 2—traditionally defining CE. Furthermore, due to experimental observations [18] of marked differences between the two yeast phenotypes with regard to glucose uptake kinetics, we consider v_o and v_e as a function of glucose uptake v_g represented by series of glucose uptake concentrations maintained "in feed" and defined by the expression $\{v_g^i \mid \forall\, i < j \quad v_g^i < v_g^j\}_{i,j=1,\dots,L}$ representing the set of constant glucose concentrations at which each simulation is run.

To formally define C⊕ we start from the observation that under this phenotype, the ratio of alcoholic fermentation over respiration increases proportionally to the glucose uptake, implying that at the maximum "in feed" glucose concentration evaluated, the ethanol secretion flux must have higher flux values with respect to the respiratory flux. In addition for the C⊕ case, we imposed that respiration and ethanol secretion should have at least one value different from zero. Formally these constraints relative to the C⊕ phenotype are summarized by logical expressions shown in Eq. 2.

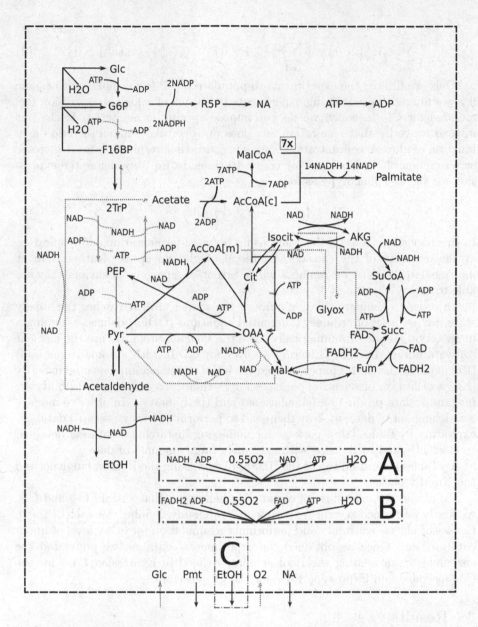

Fig. 2. Diagram of the yeast metabolism core model. The model consists in 48 reactions and 34 metabolites representing the main metabolic pathways. Only glucose is considered as carbon source. The directionality of reactions has been imposed according to literature. A and B indicate reactions modeling respiration, while C labeled reaction indicates the proxy for fermentation. Black solid arrows indicate reactions that significantly differ between C⊕ and C⊖ accordingly to the Kolmogorov-Smirnov test described in Sect. 3, red dashed arrows are not significant reactions accordingly to the same test. (Color figure online)

$$\Big(\sum_{l=1}^{L} v_e(v_g^l) > 0 \Big) \wedge \Big(\sum_{l=1}^{L} v_o(v_g^l) > 0 \Big) \wedge \Big(v_e(v_g^L) - v_o(v_g^L) > 0 \Big) \qquad (2)$$

Once we filtered the experiments to populate the C⊕ ensemble, on the same line we filtered the remaining experiments by means of a logical expression formulated for C⊖ in which we do not impose specific fermentation levels, but instead we verify that respiration flux does not overtake ethanol secretion as a function of glucose concentrations, implying also that respiration must increase as a function of "in feed" glucose concentrations. In Eq. 3 we define the expression for the C⊖ phenotype as follows:

$$\Big(v_o(v_g^1) - v_o(v_g^L) < 0 \Big) \qquad (3)$$

In this paper, we focus on the relationship between fermentation and respiration at the extremes of the considered interval of glucose uptake (rather than at intermediate levels) to examine a wider extent of emergent behaviors that are able to satisfy Eqs. 2 and 3.

In order to simulate the dynamics of the model until it reaches the steady state, we use a set of ordinary differential equations (ODEs) defined assuming a mass action kinetic. To numerically solve the ODEs system, we use the efficient software library LSODA (Livermore solver for ODEs with automatic method) [19]; in particular in this paper we used the LSODA version implemented in SciPy [12] (a collection of scientific packages for Python). To gain more flexibility we uncoupled data production (simulations) and their analysis. In order to manage a wide amount of data, to store them and to perform queries we setup a database exploiting PyTables [1], a package for managing hierarchical datasets designed to efficiently and easily cope with extremely large amounts of data.

PyTables is built on top of the HDF5 library, using the Python language and the NumPy package.

To obtain the ensembles of metabolic phenotypes that sustain C⊕ and C⊖, we firstly performed several "experiments" randomly defining, for each of them, the set of kinetic constants and performing a simulation for every level of nutrient (glucose). Once we obtained the experimental data set we populated the ensembles implementing the Boolean filter defined in Expression Eq. 2 for the C⊕ ensemble and Expression Eq. 3 for the C⊖ ensemble.

3   Results

To test the procedure on the simplified yeast model, we tossed multiple different random sets of kinetic constants, for each of them we performed 10 different simulations evenly sampling the glucose interval [0, 25] mMol, keeping the concentration constant throughout the simulation (i.e. glucose is "in feed") time of 50 s (defined accordingly to [21]).

After the simulation we checked that the system reached the steady state: we calculated the standard deviation (σ) for every species in the system during

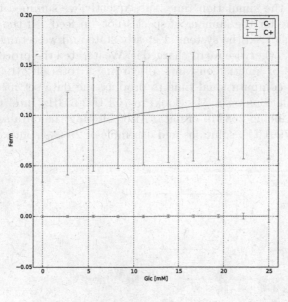

(a) Fermentation flux

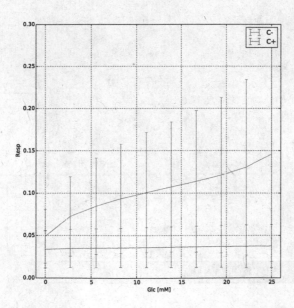

(b) Respiration flux

Fig. 3. Lines represent average flux values for C⊕ (red) and C⊖ ensembles at the variation of the glucose level, associated error bars indicate the $\pm\sigma$ values. (Color figure online)

the last 10% of the simulation time, subsequently we summed the σs and we divided the value for the number of species not "in feed". If the value was less than 1% we considered the system at steady state and we retained the random parametrization, otherwise we discharged it. We iterated the procedure to obtain 10^4 random sets of kinetic constants, discarding a total of 23199 parametrizations. The total computational time to produce the data set has been 5.5 h to run ODEs simulations on a MacBookPro (CPU 2.6 GHz Intel Core i7, RAM 16 GB) and producing 268 Mb of data. After filtering the data set we obtained an ensemble of 7901 C$\ominus$ solutions and ensemble of 29 C$\oplus$ solutions.

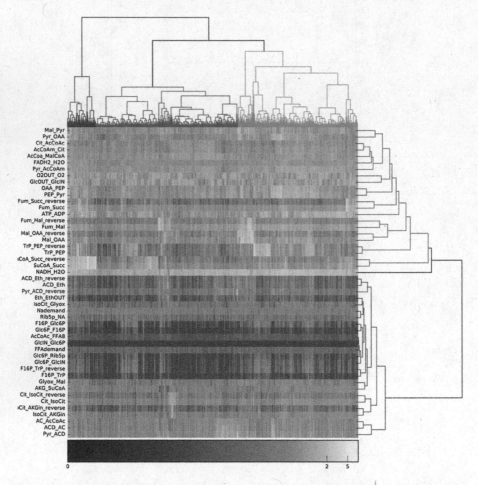

Fig. 4. Hierarchical clustering performed on both reactions (rows) and solutions (columns). The associated heatmap illustrates flux values for every reaction and for every solution at steady state. Reactions are indicated as substrate_product, with substrate and product being one among the reaction substrates and products respectively. Reverse reactions are considered separately and are indicated with the suffix "reverse". (Color figure online)

Results indicate that less than 0.3% of the total random parametrizations led to a C⊕ phenotype, while the 79% of the cases were assigned to the C⊖ ensemble.

Moreover, the 21% of the random parametrizations were not assignable to one of the two ensembles. Comparing these results with those previously published with the eeFBA approach [6] it is possible to notice that in spite of having different proportions (2% against 0.3% for C⊕ and 11% against 79% for C⊖) there still a higher probability to observe the metabolic response typical of the Crabtree-negative yeasts than of observing the Crabtree effect. The most marked difference with the eeFBA approach can be identified when comparing solutions that were not assigned to any ensemble (87% against 21%).

To improve the soundness of the biological readout, we analyzed the resulting ensembles by means of a hierarchical clustering performed on the global data set computing euclidean distances on the flux values matrix (i.e. the flux distribution in the different solutions). In particular we clustered both reactions (rows in Fig. 4) and solutions (columns in Fig. 4), overall we stress the fact that C⊕ and C⊖ solutions form two main separate clusters (C⊕ : small green cluster on columns, C⊖ : red and light blue clusters on columns) and, for what concerns reactions, on rows three main clusters are clearly evident (red, green and blue); however there is no strict correspondence between reaction clustering and biochemical pathways (e.g., reactions of both TCA cycle and glycolysis are split among two different clusters).

To further characterize the ensembles identifying those fluxes significantly different between the two ensembles we exploited a Kolmogorov-Smirnov test [13], a non-parametric hypothesis test procedure able to discriminate if two samples derive from the same distribution without investigating the actual shape of the distributions. The statistical test has been performed using the flux values of each reaction at steady state for all the 10 non-null levels of glucose.

As a result we obtained 33 reaction fluxes (over a total of 48) that are significantly different for at least 9 out of 10 levels of glucose (indicated with black solid arrows in Fig. 2), setting a p-value threshold of 0.05. Obviously among these are the reactions used to discriminate between the C⊖ and C⊕ phenotypes.

4   Conclusion

Constraint-based models have been effectively used to study metabolic fluxes at steady state, however they are not able to provide information on the temporal evolution of the system during the transient phase previous to the steady state. Moreover with constraint-based methods it is not possible to infer the metabolic concentrations at steady state due to the fact that there is no information about kinetic constants (a metabolic flux is determined using the Eq. 1).

The technique developed in the present work together with the current calculating capacity, enables to overcome this limitations by means of mechanism-based simulations (parametrized used random kinetic constants and initial molecular concentrations retrieved in literature), calculation of metabolic fluxes at steady state and selection of those solutions (sets of kinetic constants and molecular concentrations at steady state) that are in agreement with phenotype definitions. Results shown in Fig. 3, illustrates that by a simple filtering of pivotal fluxes for respiration and fermentation at the boundary levels of nutrient uptake, the devised method is able to discriminate between $C\oplus$ and $C\ominus$ metabolic phenotypes. Indeed, as it happens *in vivo*, fermentation flux is higher in simulated $C\oplus$ solutions with respect to $C\ominus$ for every level of glucose, while respiration shows an opposite behavior (the average respiration flux in $C\ominus$ ensemble is higher than in $C\oplus$). Moreover, the hierarchical clustering illustrating a separation between $C\oplus$ and $C\ominus$ solutions as well as the Kolmogorov-Smirnov test identifying as statistically different pivotal reactions for the identification of the two phenotypes, provide a further support for the obtained results.

In conclusion, in this work we provided a proof of concept for a computational framework able to discriminate between different metabolic phenotypes in order to retrieve ensembles of putative steady state metabolic concentrations and kinetic constants without the need of assuming that the cell is optimized towards a specific behavior.

Currently we are exploiting the computational framework here described to investigate the linking between alterations in metabolic fluxes and shifts in metabolite levels. Briefly, preliminary results suggest that metabolite levels exhibit a poor correlation with variations in flux values of reactions directly involving them. At the same time, results show a stronger linkage with variations in fluxes that are distantly located in the network [7].

In the next future, we plan to expand the set of sampled random kinetic constants and to implement a more efficient strategy to determine the metabolic steady state (e.g. exploiting the NLEQ2 algorithm [16]). Lastly, we will consider the feasibility of using parallel and high performance computing techniques [14,15] in order to speed-up simulations.

Acknowledgments. This work has been supported by SYSBIO Centre of Systems Biology, through the MIUR grant SysBioNet—Italian Roadmap for ESFRI Research Infrastructures.

References

1. Alted, F., Vilata, I., et al.: PyTables: hierarchical datasets in python (2002). http://www.pytables.org
2. Aung, H.W., Henry, S.A., Walker, L.P.: Revising the representation of fatty acid, glycerolipid, and glycerophospholipid metabolism in the consensus model of yeast metabolism. Ind. Biotechnol. 9(4), 215–228 (2013)
3. Alberghina, L., Westerhoff, H.V. (eds.): Systems Biology: Definitions and Perspectives. Topics in current genetics, 13th edn. Springer, Heidelberg (2005)
4. Canelas, A.B., van Gulik, V.M., Heijnen, J.J.: Determination of the cytosolic free NAD/NADH ratio in saccharomyces cerevisiae under steady-state and highly dynamic conditions. Biotechnol. Bioeng. 100(4), 734–743 (2008)

5. Cazzaniga, P., Damiani, C., Besozzi, D., Colombo, R., Nobile, M.S., Gaglio, D., Pescini, D., Molinari, S., Mauri, G., Alberghina, L., et al.: Computational strategies for a system-level understanding of metabolism. Metabolites **4**(4), 1034–1087 (2014)
6. Damiani, C., Pescini, D., Colombo, R., Molinari, S., Alberghina, L., Vanoni, M., Mauri, G.: An ensemble evolutionary constraint-based approach to understand the emergence of metabolic phenotypes. Nat. Comput. **13**(3), 321–331 (2014)
7. Damiani, C., Colombo, R., Di Filippo, M., Pescini, D., Mauri, G.: Linking alterations in metabolic fluxes with shifts in metabolite levels by means of kinetic modeling. In: Rossi, F., Piotto, S., Concilio, S. (eds.) WIVACE 2016. CCIS, vol. 708, pp. 138–148. Springer, Cham (2017). doi:10.1007/978-3-319-57711-1_12
8. De Deken, R.H.: The crabtree effect: a regulatory system in yeast. J. Gen. Microbiol. **44**(2), 149–156 (1966)
9. Di Filippo, M.: Zooming-in on cancer metabolic rewiring with tissue specic constraint-based models. Comput. Biol. Chem. **62**, 60–69 (2016)
10. Feist, A., Palsson, B.: The biomass objective function. Curr. Opin. Microbiol. **13**(3), 344–349 (2010)
11. Gianchandani, E.P., Chavali, A.K., Papin, J.A.: The application of flux balance analysis in systems biology. Wiley Interdisc. Rev. Syst. Biol. Med. **2**(3), 372–382 (2010)
12. Jones, E., Oliphant, E., Peterson, P., et al.: SciPy: open source scientific tools for python (2001). http://www.scipy.org/
13. MacFarland, T.W.W., Yates, J.M.M.: Introduction to Nonparametric Statistics for the Biological Sciences Using R. Springer, Cham (2016)
14. Nobile, M.S., Besozzi, D., Cazzaniga, P., et al.: GPU-accelerated simulations of mass-action kinetics models with cupSODA. J Supercomput **69**(1), 17–24 (2014)
15. Nobile, M.S., Cazzaniga, P., Besozzi, D., et al.: CuTtauLeaping: a GPU-powered tau-leaping stochastic simulator for massive parallel analyses of biological systems. PLoS ONE **9**(3), e91963 (2014)
16. Olivier, B.G., Rohwer, J.M., Hofmeyr, J.H.S.: Modelling cellular systems with PySCeS. Bioinformatics **21**(4), 560–561 (2005)
17. Orth, J.D., Thiele, I., Palsson, B.O.: What is flux balance analysis? Nat. Biotechnol. **28**(3), 245–248 (2010)
18. Papini, M., Nookaew, I., Uhlén, M., Nielsen, J.: Scheffersomyces stipitis: a comparative systems biology study with the crabtree positive yeast saccharomyces cerevisiae. Microb. Cell Fact. **11**, 136 (2012)
19. Petzold, L.: Automatic selection of methods for solving stiff and nonstiff systems of ordinary differential equations. SIAM J. Sci. Stat. Comp. **1**(4), 136–148 (1983)
20. Smallbone, K., et al.: A model of yeast glycolysis based on a consistent kinetic characterisation of all its enzymes. FEBS Lett. **587**(17), 2832–2841 (2013)
21. Theobald, U., Mailinger, W., Baltes, M., Rizzi, M., Reuss, M.: In vivo analysis of metabolic dynamics in saccharomyces cerevisiae: I experimental observations. Biotechnol. Bioeng. **55**(2), 305–316 (1997)
22. Zhao, J., Yu, H., Luo, J., Cao, Z., Li, Y.: Complex networks theory for analyzing metabolic networks. Chin. Sci. Bull. **51**(13), 1529–1537 (2006)

Process Algebra with Layers: Multi-scale Integration Modelling Applied to Cancer Therapy

Erin Scott[1] , James Nicol[2], Jonathan Coulter[2] , Andrew Hoyle[1] ,
and Carron Shankland[1(✉)]

[1] Computing Science and Mathematics,
University of Stirling, Stirling FK9 4LA, UK
ces@cs.stir.ac.uk
[2] School of Pharmacy,
Queen's University Belfast, Belfast BT7 1NN, UK

Abstract. We present a novel Process Algebra designed for multi-scale integration modelling: Process Algebra with Layers (PAL). The unique feature of PAL is the modularisation of scale into integrated layers: Object and Population. An Object can represent a molecule, organelle, cell, tissue, organ or any organism. Populations hold specific types of Object, for example, life stages, cell phases and infectious states. The syntax and semantics of this novel language are presented. A PAL model of the multi-scale system of cell growth and damage from cancer treatment is given. This model allows the analysis of different scales of the system. The Object and Population levels give insight into the length of a cell cycle and cell population growth respectively. The PAL model results are compared to wet laboratory survival fractions of cells given different doses of radiation treatment [1]. This comparison shows how PAL can be used to aid in investigations of cancer treatment in systems biology.

Keywords: Systems biology · Formal methods · Mathematical modelling · Cell cycle · DNA damage

1 Scientific Background

Multi-scale modelling in systems biology is now commonplace and indeed essential to many investigations [2,3]. Analysis of emergent properties arising from the interactions between scales of multi-scale systems is important as an aid in solutions to topical issues such as disease and climate change. Indeed the issue of cancer cell growth and damage from treatments is a multi-scale system as the damage affects the levels of intracellular proteins within a cell and therefore affects the cell population levels [1,4]. There are a number of successful multi-scale models such as Powathil et al. [4] cellular automaton model to study the dynamics of chemotherapy drugs to cancer cell-cycle heterogeneity.

© Springer International Publishing AG 2017
A. Bracciali et al. (Eds.): CIBB 2016, LNBI 10477, pp. 118–133, 2017.
DOI: 10.1007/978-3-319-67834-4_10

Another example is the Met Office climate prediction model [5] which couples the atmospheric and oceanic scales together in one model.

There is no universally adopted theoretical/computational framework or language for the construction of multi-scale models. Most multi-scale models are specific to the problem they are addressing and are defined by integrated scales that are modelled in different mathematical and computational languages [2,3]. These hybrid models use a combination of modelling approaches such as Ordinary Differential Equations (ODE) and Cellular automata (CA) to define specific scales of the model. For example, the Powathil et al. [4] model is hybrid because the intracellular proteins are defined in ODE and the cell populations are modelled in cellular automaton. The Met Office climate prediction model [5] is hybrid as it utilises different mathematical approaches to describe the different scales and is specific to the problem of climate change forecasting. These hybrid models make the structure and the analysis of the model difficult as the scales are defined in separate models. The modeller must create, or be knowledgeable in, integration techniques to link the models together.

Process algebra offers an ideal opportunity in systems biology [6]. It gives a high-level description of interactions, communications, and synchronizations between a collection of independent agents or processes. Its application provides many analysis techniques for systems' behaviour and properties. For example, time series simulations (to produce model predictions to compare with observed data), Markovian analysis (deriving a Continuous Time Markov Chain (CTMC) of all the possible states and evolutions of the model to be used for functional verification), model checking (to validate the model against a high level property specified in e.g. temporal logic), and model generation (creating models from time series data). The multiscale P-system framework of Romero-Campero et al. [7] is also attractive for its ability to describe extremely abstract hierarchical systems within one formalism; however, the range of possible analyses is smaller.

There are only a few process algebra languages that are specifically designed for multi-scale systems. These include Parametric Stochastic Process Algebra with Hooks (psPAH) [8] and Performance Evaluation Process Algebra nets (PEPA nets) [9]. These multi-scale languages focus on the integration of spatial scales, assuming the same time scale. One important multi-scale modelling feature is allowing the easy definition of the addition and deletion of objects within the language to capture, for example, cell division and cell death. Neither psPAH nor PEPA nets include this specific modelling feature. As a result, the modeller needs to add many lines of code to add and delete objects, making the model difficult to construct and read. In this paper we propose a novel language, Process Algebra with Layers (PAL), which gives a convenient representation of multi-scale systems by putting these features directly into the syntax and semantics.

The unique features of PAL are the integrated layers: Object and Population. These novel layers allow PAL to include the easy definition of the addition and deletion of Objects through Population actions. These layers modularise the definitions of specific Object populations from the Object's internal system definition giving a more elegant model. The layers are generally applicable to a variety of multi-scale systems: see Sect. 2.1. We show how easily the layers of PAL can be applied to a novel mammalian cell cycle and DNA damage case study. The case study has two distinct layers. At one level, there is a population of growing and dividing cells. How cells move between these states is controlled by changes in cell mass and levels of selected proteins. The intracellular species and interactions between them form the second layer. The novel PAL model links together the established models of Zhang et al. [10] and Tyson et al. [11] to investigate the effect of DNA damage from radiation on the progression of the cell cycle. This has not been previously considered in the literature.

2   Materials and Methods

2.1   Process Algebra with Layers

PAL has two layers which are named Object and Population. Figure 1 shows a conceptual schematic of these layers.

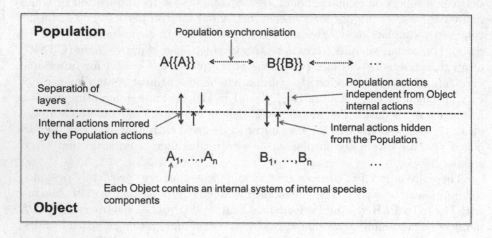

Fig. 1. Schematic of Object and Population Layers. The Object layer describes each specific Object. Objects have an internal system description of internal species components. The Population layer defines populations of Objects.

An Object is an individual system model at the lowest scale of interest. Objects comprise a number of internal species components. A PAL model may have multiple Object types, and multiple replications of these Objects. The Objects' internal species evolve dynamically over time via actions. Some actions are internal to the Object, and some impact on the next scale up, the Population. A Population is a collection of Objects. A PAL model must have at least one Population. There may be interaction between Populations and between layers.

Objects and Populations can stand for any scale the modeller chooses. In Sect. 2.5, for example, PAL is illustrated by a mammalian cell case study. The cells and their intracellular proteins (internal species) are the Objects. These drive the cell cycle. Populations define cells in specific phases such as growing and dividing. Mirrored actions connect Objects (e.g. when a protein reaches a specific level) with Populations (e.g. resulting in a state change from growing to dividing). Another example, from Chap. 5 of Scott [12], takes marine organisms as Objects, with their physiology in energy budgets as internal species. The organisms are held in Populations describing their life stage such as larvae, juvenile and adult. The organisms change life stage through mirrored actions (e.g. when mass reaches a certain threshold, larvae evolve to juveniles) and also can be removed from the system (e.g. fishing of the Population). Also at the organism level, we could consider a model of disease hosts (Objects) and their immune system and micro-parasite interaction (internal species). Populations of hosts are defined as susceptible, infected and recovered. Hosts change infectious states dependent on the number of parasites in their system (mirrored actions) and more hosts can be added due to immigration (independent Population action).

2.2   The Syntax of PAL

The syntax of PAL is shown in Fig. 2. PAL uses the same syntax as Biochemical-Performance Evaluation Process Algebra (Bio-PEPA) [13] to define the internal species components.

The component O, called an **Object component**, describes an internal system and the interactions among internal species components S. Species are named using $C = S$ to allow modular construction. The element x is a positive integer-valued parameter. Constants allow names to be assigned to patterns of behaviour associated with components. (α, κ) is the internal species prefix, where $\alpha \in$ SpeciesActions is the action type, k is the stoichiometry coefficient of the species in that reaction, and SpeciesActions is a modeller-defined set of action names. The prefix combinators op are: $<<$ indicating a reactant, $>>$ a product, $(+)$ an activator, $(-)$ an inhibitor and $(.)$ a generic modifier. $O \bowtie O$ denotes the cooperation between internal species over the cooperation set L. Set L determines those activities on which the cooperands are forced to synchronise.

$$P\{\{O\}\}_A ::= (\alpha, 1)\ PAL_{op}\ P\{\{O\}\}_A \mid P\{\{O\}\}_A\ +\ P\{\{O\}\}_A \mid D$$

Where $PAL_{op} = \downarrow \mid \uparrow \mid ((+))$

$$O ::= O \underset{L}{\bowtie} O \mid S(x)$$

$$S ::= (\alpha, \kappa)\ op\ S \mid S + S \mid C$$

Where $op = << \mid >> \mid (+) \mid (-) \mid (.)$

$$M ::= M \underset{L_s}{\Diamond} M \mid P\{\{O\}\}_A$$

Fig. 2. Syntax of PAL.

The component $P\{\{O\}\}_A$ is called a **Population component** and represents a multi-set of Object components $P\{\{O\}\}_A = P\{\llbracket O_1, ..., O_n \rrbracket\}_A$. A multiset is an unordered collection of Objects with repetitions. Populations are named using $D = P\{\{O\}\}_A$ to allow modular construction. $(\alpha, 1)$ is the prefix, where $\alpha \in$ Actions is the action type and 1 is the stoichiometric coefficient of the Object in that action. The design choice of a stoichiometry of 1 was chosen to simplify the resulting states the action produces. There are three prefix combinators called PAL_{op} which represent the role of the Objects in the action. These are: $\downarrow$ indicates a deletion of an Object, $\uparrow$ an addition of an Object (an *initialO* element will be added to a specific Population with a specific initial set up relevant to its Population) and $((+))$ an Object which is an activator (the Object is involved in the action but does not change).

The choice operator $P\{\{O\}\}_A + P\{\{O\}\}_A$ and $S + S$ represents nondeterministic choice between actions whether these be Population actions or internal species actions. Once one branch is chosen the others are discarded, thus choice represents competition between actions depending on their rate.

The top-level component M, called a model component, describes the system and the interactions among Population components. The cooperation between Populations over the multi-scale action cooperation set L_s is expressed by $M \diamondsuit_{L_s} M$. Set L_s determines those actions on which the cooperands must synchronise. Each Population component must have a hidden action set A identifying internal species actions which are hidden from the Population component. Hidden actions should not be in the set L_s in a well defined PAL system.

2.3   The Semantics of PAL

A PAL system $\mathcal{P}$ is a septuple $\langle Pcomp, Ocomp, Scomp, F_R, \mathcal{K}, \mathcal{N}, \mathcal{M} \rangle$, where:

- $Pcomp$ is the set of definitions of Population components;
- $Ocomp$ is the set of definitions of Object components;
- $Scomp$ is the set of definitions of internal species components;
- F_R is the set of functional rate definitions;
- $\mathcal{K}$ is the set of parameter definitions;
- $\mathcal{N}$ is the set of quantities describing each internal species;
- $\mathcal{M}$ is the model component describing the system.

The definition of the Object components in $Ocomp$ must be defined in terms of the internal species components defined in $Scomp$ and for each cooperation set L_i in O, $L_i \subseteq$ SpeciesActions (O). In a well-defined PAL system each element has to satisfy the following conditions. Set $\mathcal{N}$ has to contain all the internal species components. The functional rates are well defined if each variable in their definition refers to the name of a species component in the set $\mathcal{N}$ or a constant parameter in the set $\mathcal{K}$. The definition of the internal species components in $Scomp$ must have sub-terms of the form (α, k)op S and the action types in each single component must be distinct. The definition of the Population components $Pcomp$ must be defined in sub-terms of the form $(\alpha, 1)$ PAL_{op} $P\{\{O\}\}_A$ and the prefixReac $((\alpha, k) << S)(l) \xrightarrow{(\alpha,\ [S\ :<<\ (l, \kappa)])} S(l - k)$ $k \le l \le N$

prefixProd $((\alpha, k) >> S)(l) \xrightarrow{(\alpha,\ [S\ :>>\ (l, \kappa)])} S(l + k)$ $0 \le l \le (N - k)$

prefixMod $((\alpha, k)op\ S)(l) \xrightarrow{(\alpha,\ [S\ :\ op(l, \kappa)])} S(l)$ with op $= (.), (+), (\text{-})$ and

$0 < l \le N$ if $op = (+)$, $0 \le l \le N$ otherwise

where S is the name of the species component, op is the action type, l the level κ the stoichiometry coefficient, and N the maximum level of S.

choice1 $\dfrac{S_1(l) \xrightarrow{(\alpha,w)} S_1'(l')}{(S_1 + S_2)(l) \xrightarrow{(\alpha,w)} S_1'(l')}$ choice2 $\dfrac{S_2(l) \xrightarrow{(\alpha,w)} S_2'(l')}{(S_1 + S_2)(l) \xrightarrow{(\alpha,w)} S_2'(l')}$

constant $\dfrac{S(l) \xrightarrow{(\alpha,S:\ [op(l,\ k)])} S'(l')}{C(L) \xrightarrow{(\alpha,C:\ [op(l,\ k)])} S'(l')}$ with $C = S$

coop1 $\dfrac{O_1 \xrightarrow{(\alpha,w)} O_1'}{O_1 \bowtie_L O_2 \xrightarrow{(\alpha,w)} O_1' \bowtie_L O_2}$ with $\alpha \notin L$

coop2 $\dfrac{O_2 \xrightarrow{(\alpha,w)} O_2'}{O_1 \bowtie_L O_2 \xrightarrow{(\alpha,w)} O_1 \bowtie_L O_2'}$ with $\alpha \notin L$

coop3 $\dfrac{O_1 \xrightarrow{(\alpha,w)} O_1' \quad O_2 \xrightarrow{(\alpha,w)} O_2'}{O_1 \bowtie_L O_2 \xrightarrow{(\alpha,w)} O_1' \bowtie_L O_2'}$ with $\alpha \in L$

where w is a list recording the species that participate in the reaction and L is the cooperation action set

Fig. 3. Rules for Bio-PEPA included in the semantics of PAL. These rules are presented in Ciocchetta et al. [13] and are repeated here for convenience and completeness.

action types in each single component must be distinct. The model component $\mathcal{M}$ must be defined in terms of the Population components defined in $Pcomp$ and for each cooperation set L_{si} in $\mathcal{M}$, $L_{si} \subseteq$ Actions $(\mathcal{M})$.

The rules of PAL specify Population behaviour and its relation to Object behaviour. The semantics of an Object are as in Bio-PEPA, and repeated here in Fig. 3 for convenience. Figures 4 and 5 describe how Objects and Populations influence each other and how Populations evolve, respectively. These rules collectively allow a CTMC to be defined from a PAL model.

Some Object actions are hidden from the Population level as defined in the Action Hidden Rule in Fig. 4. The modeller defines a set of hidden actions A when describing a model in PAL. These could include actions such as synthesis and degradation of intracellular proteins within a cell. Actions such as these do not change the Population layer composition of the system, therefore do not

---

Action Mirror/Hidden Rules

Internal Actions that are mirrored by Populations

$$\frac{O_i \xrightarrow{(\alpha,w)} O_i'}{P\{\{O\}\}_A \xrightarrow{(\alpha,w)} P\{\{O'\}\}_A} \quad where \; \exists \, O_i \in O \; \wedge \; \alpha \notin A$$

$$where \; O' = O \; \oplus \; O_i' \wedge P\{\{O\}\} \xrightarrow{(\alpha,w)} P\{\{O'\}\}$$

Internal Actions that are hidden from Populations

$$\frac{O_i \xrightarrow{(\alpha,w)} O_i'}{P\{\{O\}\}_A \xrightarrow{\tau} P\{\{O'\}\}_A} \quad where \; \exists \, O_i \in O \; \wedge \; \alpha \in A$$

$$where \; O' = O \; \oplus \; O_i'$$

where $\oplus$ overwrites O_i *in* O with new O_i' state, leaving the rest of O unchanged.

Fig. 4. Semantics of PAL: Action Mirror/Hidden Rules.

need to be mirrored by the Population. For example, transitions of cell proteins indirectly affect mass but are hidden from the Population layer.

Object actions that are mirrored by the Population are defined by the Action Mirror Rule in Fig. 4. For example, in the cell model, the changing mass of the cell (an internal species) will trigger the transition of the cell from the growing state to the dividing state. The changing internal action in this case has an impact on the Population view of the system. These internal actions are mirrored by Population actions which are defined by the Prefix Population Transition Rules shown at the top of Fig. 5.

There are essentially three Prefix Population Transition Rules: adding, deleting and activator. The deletion rule has two variants depending on whether the deletion is initiated from the Object or the Population level. These rules are asymmetric because when deleting an Object from a Population the rule needs to identify the specific Object that is to be deleted. For example, the deletion rule can be used for cell phase transitions and deaths of specific cells, therefore the cell must be known to the rule so that the correct cell is deleted. In the case of addition and the activator rule a specific Object does not need to be known. For example, in the addition rule a new initialisation of an Object is added to a Population. The Object in the activator rule does not need to be known by the rule as the rule does not change the Object. In the case study here this would be a dividing cell becoming two growing cells.

Lastly, potential interactions at the Population level are dealt with by the Population Transition Rules (see lower section of Fig. 5). Populations can perform actions autonomously and this allows actions such as death from a Population action to be defined in a model. This feature is not used in the case

Prefix Population Transition Rules

Adding an Object to a Population

$$((\alpha,1) \uparrow P\{\{O\}\}_A) \xrightarrow{(\alpha,w)} P\{\{O'\}\}_A \text{ where } \alpha \notin A$$
$$O' = O \cup [\![initialO]\!]$$

Deleting a specific Object from a Population

$$((\alpha,1) \downarrow P\{\{O\}\}_A) \xrightarrow{(\alpha,w)} P\{\{O'\}\}_A \text{ where } \alpha \notin A \wedge \alpha \in \text{SpeciesActions}(O)$$
$$\exists i.\ O_i \in O \wedge O_i \xrightarrow{(\alpha,w)} O'_i \wedge O = [\![O_1,...,O_n]\!] \wedge |O| \geq 1 \wedge O' = O \backslash [\![O_i]\!]$$

Activator does not increase or decrease a Population

$$((\alpha,1)((+))P\{\{O\}\}_A) \xrightarrow{(\alpha,w)} P\{\{O\}\}_A \text{ where } \alpha \notin A$$
$$O = [\![O_1,...,O_n]\!] \wedge |O| \geq 1$$

Deleting a random Object from a Population

$$((\alpha,1) \downarrow P\{\{O\}\}_A) \xrightarrow{(\alpha,w)} P\{\{O'\}\}_A \text{ where } \alpha \notin A \wedge \alpha \notin \text{SpeciesActions}(O)$$
$$\exists i.\ O_i \in O \wedge O = [\![O_1,...,O_n]\!] \wedge |O| \geq 1 \wedge O' = O \backslash [\![O_i]\!]$$

where (α,w) comes from synchronising with another Population
or from the Object layer.

Population Transition Rules

Constant

$$\frac{P\{\{O\}\}_A \xrightarrow{(\alpha,w)} P\{\{O'\}\}_A}{D \xrightarrow{(\alpha,w)} P\{\{O'\}\}_A} \text{ where } D = P\{\{O\}\}_A$$

Asynchronous Left

$$\frac{P\{\{O\}\}_A \xrightarrow{(\alpha,w)} P\{\{O'\}\}_A}{P\{\{O\}\}_A \underset{L_s}{\Diamond} M \xrightarrow{(\alpha,w)} P\{\{O'\}\}_A \underset{L_s}{\Diamond} M} \text{ where } \alpha \notin L_s$$

Asynchronous Right

$$\frac{P\{\{O\}\}_A \xrightarrow{(\alpha,w)} P\{\{O'\}\}_A}{M \underset{L_s}{\Diamond} P\{\{O\}\}_A \xrightarrow{(\alpha,w)} M \underset{L_s}{\Diamond} P\{\{O'\}\}_A} \text{ where } \alpha \notin L_s$$

Population Synchronisation

$$\frac{P_1\{\{O_1\}\}_A \xrightarrow{(\alpha,w)} P_1\{\{O'_1\}\}_A \quad P_2\{\{O_2\}\}_A \xrightarrow{(\alpha,w)} P_2\{\{O'_2\}\}_A}{P_1\{\{O_1\}\}_A \underset{L_s}{\Diamond} P_2\{\{O_2\}\}_A \xrightarrow{(\alpha,w)} P_1\{\{O'_1\}\}_A \underset{L_s}{\Diamond} P_2\{\{O'_2\}\}_A}$$

where $\alpha \in L_s$

where L_s is the multi-scale synchronisation action set

Fig. 5. Semantics of PAL: Prefix Population Transition Rules and Population Transition Rules.

study, but could be added if, for example, chemical messengers from a dying cell influenced death in another cell. Populations synchronise/communicate on specific external actions as defined in the Population Transition Rules shown at the base of Fig. 5. This allows the definition of cell phase transitions and cell division as these actions involve two Populations changing in number. For more detail on PAL see Chap. 4, pp. 74–80 of Scott [12].

2.4   Case Study

PAL is applied to a mammalian cell cycle and DNA damage case study to illustrate its capabilities in systems biology. A cell cycle is the series of events that take place in a cell leading to its division. The motivation of this case study is to analyse the effects of damage from radiation treatments to the length of a cancer cell cycle and cell survival. A PAL model has been created by linking together an established cell cycle model from Tyson et al. [11] with a repair model with an external force applying damage by Zhang et al. [10]. Other models such as Powathil et al. [4,14] and Guerrero et al. [15] use the Tyson et al. [11] model as a basis for cell cycle transitions and regulation. Zhang et al. [10] presents a number of potential models for the transcription factor p53 activity observed experimentally in response to DNA damage. p53 is at the centre of a number of DNA damage responses which interact downstream with the regulation of the cell cycle. The linking of the Zhang et al. [10] model with the Tyson et al. [11] model allows the creation of a novel model investigating the effects of DNA damage from radiation treatments on the species affecting progression of the cell cycle and consequent effect on cell colonies. This has not been previously considered in the literature. The novel PAL model allows multi-scale analysis, including Object layer experimentation (average length of a single cell cycle, Sect. 3.1) and Population layer experimentation (cell population growth, Sect. 3.2).

2.5   PAL Model

Although PAL makes describing the model simpler than other multi-scale techniques, the model is too long to be shown here. See Chap. 6, pp. 109–112 of Scott [12]. The description and results of experimentation are given here. The case study has two distinct layers: cell population and intracellular. The cell Population layer is described in the PAL model by defining two PAL Populations based on the two steady states of a cell: Growing and Dividing. These Populations contain *G cell* and *D cell* Object components, illustrated in Fig. 6 by a Growing cell becoming a Dividing cell, and that in turn becoming two Growing cells.

In the intracellular layer, G and D cells contain internal species which are the cell mass and proteins translated from Tyson et al. [11] and Zhang et al. [10]. These proteins include the Cdk-cyclin B complex (CycB), the APC-Cdh1 complex (Cdh1), the active form of Cdc20 (Cdc20A), the total Cdc20 (Cdc20T) and the intermediary enzyme (IEP). These are shown as the internal species in Fig. 6.

The graphic shows how the proteins rise and fall in response to each other, creating the conditions of the cell cycle. Note the black label in the graphic indicating Growing phase or Dividing phase.

Transitions between the two Populations are controlled by changes in cell mass and threshold values of the CycB, indicated on the arrows in Fig. 6. To make the cell cycle relevant to mammalian cells the parameter values of this model are taken from Powathil et al. [4], therefore, time in the model is in hours.

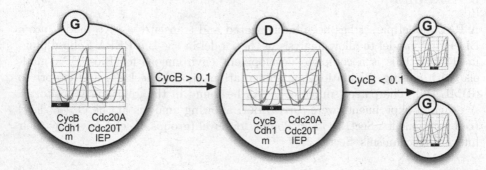

Fig. 6. Example of a single G cell evolving, through its internal species, to a D cell, and then to two G cells.

To model how the cell cycle proteins are affected by DNA damage a ODE model originally developed by Zhang et al. [10] is translated into internal species and parameters of the PAL model. The model consists of DNA damage caused by radiation treatment, the p53 and Mdm2 (nucleus and cytoplasmic) that promote the degradation of p53. p53 inhibits the activity of CycB preventing the progression of the cell cycle. In the PAL model the Tyson et al. [11] and Zhang et al. [10] models are innovatively linked together by changing the CycB degradation rate to be influenced by changes in the p53 levels. Levels of p53 are at equilibrium when there is no damage in the system. When there is damage it causes p53 levels to pulse according to Zhang et al. [10]. Damage is a parameter of the model and Zhang et al. [10] states the simple assumption is made that damage is repaired at a constant rate.

Damage is not uniform: although the whole plate of cells has the same treatment, radiation at lower levels will hit some cells but not others. To model this, different damage levels are assigned to each cell in the colony depending on the highest level of damage at the start of the simulation. For example, if the highest damage in a simulation is five the cells are assigned damages in the range of zero to five on a random distribution. Damage occurs immediately in all simulation experiments and ranges from 0–12 (integer values). We assume damage greater than four causes cell death through the Population actions. Cell cycle length is impacted with damage of four or less by the increase of the CycB degradation rate.

The investigation of the effects of damage in the PAL model from the average length of a cell cycle (Object layer) to population growth (Population layer) was carried out. The damage in the model is abstract, therefore matching the real notion of damage from radiation was achieved by the creation of a relationship function between the abstract damage and Gy, $damage/2 = $ Gy dose, chosen here to fit with experimental results, see Sect. 3.3. This is why damage ranges from 0–12, to fit with 0 to 6 Gy.

3   Results

A PAL model parser has been implemented and translates a PAL model into a Bio-PEPA model to allow analysis of the model in the Bio-PEPA Eclipse plug-in [13,16]. This is a complete development environment for Bio-PEPA models, with editing, simulation, experimentation, model checking, and export to SBML [17]. The parser source code can be found in the following repository[1]. A range of experiments were carried out, allowing analysis at the Object level (cell components, Sect. 3.1) and Population level (groups of cells driven by their internal mechanisms, Sect. 3.2).

3.1   Object Layer Experiments: Analysis of Average Length of a Cell Cycle

Simulation distribution analysis was undertaken to analyse the average length of the cell cycle and the impact of increasing the amount of damage. This analysis takes place at the individual cell scale (Object layer). Simulation distributions obtain the percentage of a user-defined number of stochastic simulations for which some property is true at or before a given time t. The Bio-PEPA plug-in plots the Cumulative Distribution Function (CDF) and Probability Distribution Function (PDF) of any agents in the model, with respect to the target value.

Table 1. Average length of cell cycle and 95% confidence interval in hours of each simulation distribution experiment.

Experiment	Average cycle	Confidence interval (95%)
Control	23.96	(23.51, 24.41)
Damage 1	24.18	(23.70, 24.66)
Damage 2	24.73	(24.16, 25.30)
Damage 3	24.74	(24.09, 25.39)
Damage 4	25.88	(25.19, 26.57)

Five experiments were carried out (damage 0–4, i.e. no cell death), see Table 1 for results. As this analysis is observing one cell cycle, the PAL model starts

---

[1] PAL Parser source code: https://github.com/MissErinScott/PAL-Parser.

with one G cell with the specified level of damage. The chosen component in this analysis is an agent which tracks a cell's completion of one cell cycle. The number of stochastic simulation replications is 200 and the stop time is 48 h. The computation time for each experiment was approximately 3 h (MacOS X Yosemite version 10.10.5, 2.2 GHz Intel core i7, 16 GB 1600 MHz DDR3).

All simulations completed a cell cycle before the stop time of 48 h. The results show that damage from one to three does not significantly affect the cell cycle average length. The intracellular proteins can cope with these damage levels. The average cell cycle length increases when damage of four is applied.

3.2   Population Layer Experiments: Analysis of Cell Population Growth

Discrete stochastic simulation time-series analysis was carried out to analyse cell growth over a longer period and the effects of damage on a colony of cells. This analysis takes place at the colony scale (Population layer).   The initial population was eight G cells (due to population limitations of Bio-PEPA plug-in [13]). Different damage levels are assigned to each cell randomly depending on the highest level of damage at the start of the simulation. Cells are simulated from 0–64 h. Throughout the time period of the simulation, new cells will be assigned different damage levels based on a constant repair rate. Experiments were carried out with damage ranging from 0 (control) to 12. Results from one replication are presented in Fig. 7 which shows the total population growth of G and D cells at different damage levels. Each experiment is one stochastic simulation which had a computation time of 3 to 20 min dependent on damage and a further 15 min for manual processing. Four replications of these stochastic simulations were carried out for damage values 0 to 6, 9 and 12.

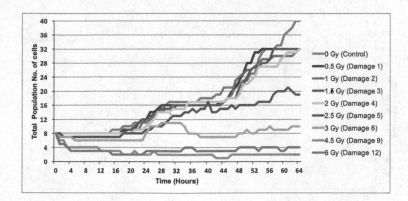

Fig. 7. Single replication of total cell population growth at different damage levels.

The results show that population growth is not affected by damage from one to four, as expected. These results reinforce the simulation distribution results which show the cell cycle is not affected by this lower damage range. The damage in each experiment is repaired at a constant rate, therefore the population of cells starts to recover, dependent on the damage assigned. Death occurred in the experiments where damage is above four. This feature had the effect of reducing the population substantially. The populations assigned the damage levels of nine and twelve had a greater reduction as they had the greater proportion of cells assigned a damage above four.

3.3 Comparison with Wet Laboratory Survival Fraction Results

The population results of the PAL model are compared to wet laboratory survival fraction results of cells given different Gy doses of radiation treatment [1]. This data is shown in Fig. 8 (red squares) and was generated using Clonogenic survival assays following treatment with 0–6 Gray (Gy) 160 kVp x-rays as previously described by Butterworth et al. [18]. The survival fractions were calculated as the plating efficiency of the treated group divided by the plating efficiency of the untreated control cells, with error bars representing the standard deviation (SD) (n = 7). The wet laboratory experiments had duration of twelve days and initial populations of 200 to 600 cells. Previously, Butterworth et al. [18] showed that all damage to cells would take place within 48 h (2 days); however, twelve days are required for observable colonies to form. The advantage of computational modelling is that the results can be observed at 48 h and assumptions made that if cells have survived to 48 h then they will form colonies by 12 days. We add a margin of 16 h to be sure all damage is accounted for. The complexity of the model (each cell has 21 internal species and 31 actions) mean that the evolution of an initial population of 8 G cells can be computed in reasonable time (3–20 min for each simulation, as in Sect. 3.2). In 64 h, these will grow to at most 35 G and 35 D cells; a potential total of 1470 species.

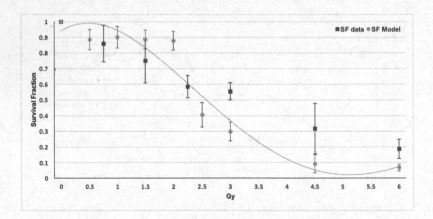

Fig. 8. Comparison results between PAL model (blue circles) and radiation treatment (red squares) survival fractions. Error bars give ± SD and fitted line to model (dotted line). (Color figure online)

For each of the four replications, the end population value data point is taken from each population experiment at different damage levels as in Fig. 7 and a survival fraction is calculated based on the control experiment. The mean survival fraction of the four replications is shown in Fig. 8 with error bars representing the standard deviation (n = 4).

The survival fraction results are compared with the damage levels based on the simple relationship $damage/2$ = Gy dose, based on the approximate alignment of 3 Gy with damage 6. Damage levels 0, 3, 6, 9 and 12 are compared with radiation doses of 0, 1.5, 3, 4.5 and 6 respectively.

Based on these simple assumptions, the results show that the model gives a closer fit to lower Gy doses (high cell survival) but an overestimation of death at higher dose levels (low cell survival). Clearly, more needs to be done to refine the model, but the point here is to illustrate the utility of PAL.

4 Conclusion

In this paper we have discussed the definition of Process Algebra with Layers (PAL), a multi-scale process algebra designed to model multi-scale systems. PAL's strength is that it allows the convenient representation of a multi-scale system in one model, in contrast to the current hybrid frameworks. PAL removes the need for the modeller to focus on the integration of the separate modelling languages that define the separate scales in a hybrid model. The novel features of PAL are the layers of the language: Object and Population. These layers allow the user to elegantly describe the the components of each scale and the interactions between scales in one PAL model. This can allow mechanistic models to be developed showing how one layer affects another.

The addition and deletion of objects is a feature of many multi-scale systems. For example, in this study cell division requires addition and cell death requires deletion. The syntax and semantics of PAL allows this feature to be easily defined by a single action integrating the scales. In comparison, existing multi-scale Process Algebra languages psPAH [8] and PEPA nets [9] would need multiple lines of code to define this feature, making their models difficult to construct and read.

The Objects in PAL currently do not have the ability to interact explicitly with one another, which may be a limitation. This would involve explicitly modelling space. Objects would, for example, need location attributes to react to their surrounding Objects. It would be necessary to ensure this addition would not compromise the integrative nature of PAL. This spatial definition may overcomplicate the definition of a PAL model which may lead to the loss of some multi-scale features PAL already encapsulates.

PAL has been applied successfully to a cell cycle and DNA damage multi-scale system here. The PAL model links together the established models of Zhang et al. [10] and Tyson et al. [11]. This allowed the creation of a novel model investigating the effects of DNA damage from radiation on a cell colony by linking mechanistically to the progression of the cell cycle as determined by cell proteins and mass. PAL easily captures the internal species and the colony activity, thus supporting investigation across scales in one model. The Object and Population results of the model showed that low radiation doses do not significantly affect the cell cycle average length, nor do they significantly affect colony growth of cancer cells.

The comparison of results to wet laboratory data shows how PAL can be used to aid in investigations of cancer treatment in systems biology. For example, the profile of Fig. 8 suggests the model can be further analysed by varying the threshold for cell death, and by modifying the simple assumption of $damage/2 =$ Gy. The model could also be refined by including more varied notions of repair, which would need targetted wet lab experiments to measure repair rates. Exploration of hypotheses in the PAL model develop understanding of the system and direct attention to the most sensitive areas for parameters or compounds. This, in turn, allows researchers to develop a more focussed programme of future biological experiments in DNA damage, cell cycle and population growth rates, reducing the number of expensive and time-consuming biological wet laboratory work. The strength of modelling approaches can be directly correlated with how they can affect broader science questions in a multi-disciplinary approach.

Future work on this PAL model could include testing a variety of different degradation rates for CycB affected by p53. The damage repair could be more specific to the levels of p53 and Mdm2. Further work could be undertaken to compare the PAL model results to other cancer treatments such as Temozolomide (TMZ) and combination of these treatments (radiation + TMZ). This comparison could be achieved easily as the damage is abstract in the model therefore the focus can be made on the damage relationship function to the specific treatment. Future work on PAL itself will include a direct implementation of a PAL tool (thereby avoiding the limitations of translation into Bio-PEPA), exploration of other analysis techniques such as model checking, and translation to/from common languages such as SBML (already available for Bio-PEPA).

Acknowledgements. Erin Scott is grateful to the Scottish Informatics and Computer Science Alliance (SICSA), a research initiative of the Scottish Funding Council, for financial support of her Ph.D. studies. We wish to thank the EPSRC project EP/K039342/1 for supplying us with the idea of the case study and the wet laboratory data. We also thank the anonymous referees for their helpful comments in improving this document.

References

1. Nicol, J.: Radiation dose enhancement: the development and application of radio consisting gold nanoparticles. Ph.D. thesis, Queen's University Belfast (2016)
2. Dada, J.O., Mendes, P.: Multi-scale modelling and simulation in systems biology. Integr. Biol. **3**(2), 86–96 (2011). Quantitative biosciences from nano to macro
3. Walker, D.C., Southgate, J.: The virtual cell-a candidate co-ordinator for 'middle-out' modelling of biological systems. Brief. Bioinform. **10**(4), 450–461 (2009)

4. Powathil, G.G., Gordon, K.E., Hill, L.A., Chaplain, M.A.J.: Modelling the effects of cell-cycle heterogeneity on the response of a solid tumour to chemotherapy: biological insights from a hybrid multiscale cellular automaton model. J. Theor. Biol. **308**, 1–19 (2012)
5. Johns, T.C., Carnell, R.E., Crossley, J.F., Gregory, J.M., Mitchell, J.F.B., Senior, C.A., Tett, S.F.B., Wood, R.A.: Low: the second Hadley centre coupled ocean-atmosphere GCM: model description, spinup and validation. Clim. Dyn. **13**(2), 103–134 (1997)
6. Priami, C.: Process calculi and life science. Electron. Notes Theor. Comput. Sci. **162**, 301–304 (2006)
7. Romero-Campero, F.J., Twycross, J., Cao, H., Blakes, J., Krasnogor, N.: A multi-scale modeling framework based on P systems. In: Corne, D.W., Frisco, P., Păun, G., Rozenberg, G., Salomaa, A. (eds.) WMC 2008. LNCS, vol. 5391, pp. 63–77. Springer, Heidelberg (2009). doi:10.1007/978-3-540-95885-7_5
8. Degasperi, A.: Multi-scale modelling of biological systems in process algebra with multi-way synchronisation. Ph.D. thesis, University of Glasgow (2011)
9. Gilmore, S., Hillston, J., Ribaudo, M.: PEPA nets: a structured performance modelling formalism. In: Field, T., Harrison, P.G., Bradley, J., Harder, U. (eds.) TOOLS 2002. LNCS, vol. 2324, pp. 111–130. Springer, Heidelberg (2002). doi:10.1007/3-540-46029-2_7
10. Zhang, T., Brazhnik, P., Tyson, J.J.: Exploring mechanisms of the DNA-damage response: p53 pulses and their possible relevance to apoptosis. Cell Cycle **6**(1), 85–94 (2007)
11. Tyson, J.J., Novak, B.: Regulation of the eukaryotic cell cycle: molecular antagonism, hysteresis, and irreversible transitions. J. Theor. Biol. **210**(2), 249–263 (2001)
12. Scott, E.: Process algebra with layers: a language for multi-scale integration modelling. Ph.D. thesis, University of Stirling (2016). http://hdl.handle.net/1893/23516
13. Ciocchetta, F., Hillston, J.: Bio-PEPA: a framework for the modelling and analysis of biological systems. Theor. Comput. Sci. **410**, 3065–3084 (2009)
14. Powathil, G.G., Adamson, D.J., Chaplain, M.A.: Towards predicting the response of a solid tumour to chemotherapy and radiotherapy treatments: clinical insights from a computational model. PLOS Comput. Biol. **9**(7), e1003120 (2013)
15. Guerrero, P., Alarcón, T.: Stochastic multiscale models of cell population dynamics: asymptotic and numerical methods. Math. Model. Nat. Phenom. **10**(1), 64–93 (2015)
16. Eclipse Foundation: Eclipse (2017). https://eclipse.org/
17. SBML.org: SBML (2012). http://sbml.org/Main_Page
18. Butterworth, K.T., Nicol, J.R., Ghita, M., Rosa, S., Chaudhary, P., McGarry, C.K., McGarry, H.O., Jimenez-Sanchez, G., Bazzi, R., Roux, S., Tillement, O., Coulter, J.A., Prise, K.M.: Preclincial evaluation of Gold-DTDTPA nanoparticles as theranostic agents in prostate cancer radiotherapy. Nanomedicine **11**, 2035–2047 (2016). doi:10.2217/nnm-2016-0062

A Problem-Driven Approach for Building a Bioinformatics GraphDB

Antonino Fiannaca, Massimo La Rosa(⊠), Laura La Paglia, Antonio Messina,
Riccardo Rizzo, and Alfonso Urso

ICAR-CNR, National Research Council of Italy,
via Ugo La Malfa 153, 90146 Palermo, Italy
{antonino.fiannaca,massimo.larosa,laura.lapaglia,
antonio.messina,riccardo.rizzo,alfonso.urso}@icar.cnr.it

Abstract. The development of high throughput technology in biological
and medical domains has seen a growing intervention of informatics support. Indeed, the big amount of data produced is difficult to analyse and
interpret in terms of time consuming and number of different resources
used. In this context, the challenge would be to have an integrated and
multi component database with a user friendly interface able to solve biological problems without a priori high-level of bioinformatics knowledge.
This need arises from the evidence that biologists have multi-task and
multi-levels problems to solve. To this aim, we propose a bottom-up,
graph-based approach for integrating bioinformatics resources, usually
databases, starting from typical biological scenarios, in order to solve
novel bioinformatics problems. The integrated resources can be queried
by means of a graph traversal language such as Gremlin.

1   Introduction

In the era of "big data" and "next generation" technologies, the role played by
bioinformatics resources is becoming central in solving biological tasks. Indeed,
the "problem solving" activity of biologists is always more complex and it needs
to manually integrate many heterogeneous publicly available resources. The
exploration and analysis of these data, also requires some knowledge on the
use of specialized tools and web servers, and the capability to move through
different services and different web interfaces; this is time-consuming in terms of
transferring data from a resource to another one, sometimes with issues related
to different aliases and accession IDs. In this context, the challenge is to provide a shared user-friendly platform hosting many bioinformatics resources, that
allows the users to solve problems without a priori high-level programming and
scripting languages knowledge [17]. Some efforts have been made to integrate,
manage, mine and do comparative analysis of high throughput data in this last
decade, as the development of bioinformatics enrichment tools for the functional
analysis of large list of genes [14], target analysis tools [9], tools for the study
of protein motifs linked to cellular pathways [24]. However, an integrated and
multi component database for bioinformatics research is still a goal to reach.

A. Bracciali et al. (Eds.): CIBB 2016, LNBI 10477, pp. 134–144, 2017.
DOI: 10.1007/978-3-319-67834-4_11

A typical example of the tedious work for a biologist in bioinformatics analysis is the enrichment analysis of microRNA (miRNA) target: starting from a list of miRNA there are different resources used as miRNA-target interaction tools that identify the targets of the selected miRNAs; these targets represent the input of other services for enrichment analysis and pathway analysis, and this implies the use of web resources completely different from each other, in addition to the time-consuming analysis of data.

In this paper, we propose a bottom-up, graph-based approach for integrating bioinformatics resources, typically databases, in order to solve novel bioinformatics problems. The integration and connection of different databases and web resources would allow the user: (1) to access different independent analysis tools, (2) to move rapidly through the resources using a common and easy interface and query language, (3) to move dynamically in the network of services and (4) to create a knowledge base (KB) to explore and solve other problems not evidenced before.

2 Background

In this section, we present some publicly available resources in the field of bioinformatics. They are widely used in order to face specific tasks, such as protein function and gene enrichment analysis.

Gene-related resources. The NCBI Entrez Gene database [23] is one of the most complete repository for genes of several species. It collects information about ortholog and homolog genes, genomic context, interactions among genes, and so on. The UCSC Genome Bioinformatics is another platform based on genome sequence data integrated with a large collection of aligned annotations [22]. The Hugo Gene Nomenclature Committee (HGNC) is a resource dedicated to human gene nomenclatures. This db contains also information about synonyms for each gene and corresponding IDs of other gene databases [13].

Protein-related resources. The UniProt Knowledgebase (UniProtKB) is the richest public repository of sequence informations and annotations about proteins. Its entries are both computationally analysed and manually annotated [27]. Other web tools allow to visualize known and predicted protein-protein interactions as the Search Tool for the Retrieval of Interacting Genes/Proteins (STRING) [25]. It includes both direct (physical) and indirect (functional) associations. Among integrated tools for protein analysis there is the Protein Analysis Through Evolutionary Relationships (PANTHER). It is a resource for comprehensive protein evolutionary and functional classification, including tools for large-scale biological data analysis [19].

Annotation-related resources. The Gene Ontology (GO) Consortium is a web resource for genes and proteins annotation [26]. It is divided in three main categories: biological processes, cellular components and molecular functions [5]. The Database for Annotation, Visualization and Integrated Discovery (DAVID) gives functional interpretation of large lists of genes [8]. It has five integrated functional annotation tool suites.

MiRNA/ncRNA-related resources. The microRNA database (miRBase) is a miRNA database containing miRNA sequences of precursor and mature forms, It gives informations about ID, genomic location and annotations [16]. MirWalk is a web archive, supplying a very large collection of predicted and experimentally verified miRNA-target interactions [9]. Moreover non-coding RNA (ncRNA)-miRNA and miRNA-miRNA interaction can be visualized. The miR-Ontology database (miRò) [18] integrates data about miRNAs, miRNA-target interactions, functional annotations provided by GO and gene-disease relations. PolymirTS is a db that supplies information about snp (single nucleotide polymorphism) in miRNA seeds and in gene targets [4] allowing to study the effect of these mutations on the binding site of miRNA targets.

Pathway-related resources. KEGG Pathway [15] is a part of the integrated DB embracing also genes, genomes, orthology, functional annotations, compounds, reactions, diseases and drugs. Reactome is an integrated db that provides validated metabolic pathways, including annotations about genes and proteins involved. It is also enriched with functional annotation functions [7].

Disease-related resources. The OMIM system is probably the most comprehensive catalogue of human genes, genetic disorders and traits, focusing mainly in gene-phenotype relationship [1]. Other resources linked to disease have more specific focuses as miRCancer, a db collecting miRNA expression profiles in different human cancer types, which are automatically extracted from literature [28].

3   Method

We propose a bottom-up approach: starting from a set of N problems, we incrementally populate our knowledge base, in the form of a graphDB, with all the resources that are necessary to solve these problems, in order to obtain a framework that allows to solve a set of M problems, with $M \geq N$. The population of the graphDB is done through the implementation of a set of customized Extract-Transformer-Loader (ETL) modules. An ETL is a computer program that parses the data files related to each resource and import them into a graphDB. These data are arranged in a set of nodes (or vertices) and edges. Each biological entity and its properties are modelled with a node and its attributes. Relationships between two biological entities are modelled by means of an edge between the two corresponding nodes. If a relation has some properties, they can be imported as edge's attributes. For example, the interaction between a microRNA (miRNA) and its target gene can be represented by means of an edge that links the node representing that miRNA and the node representing that gene. The properties of the interaction relation, such as the locus of the target site or the free energy value, are inserted as edge's attributes.

The whole population procedure can be summarized by the following four main steps.

The first step is to analyse the first problem P_1 and to identify which publicly available resources (opportunely combined) are able to face this problem.

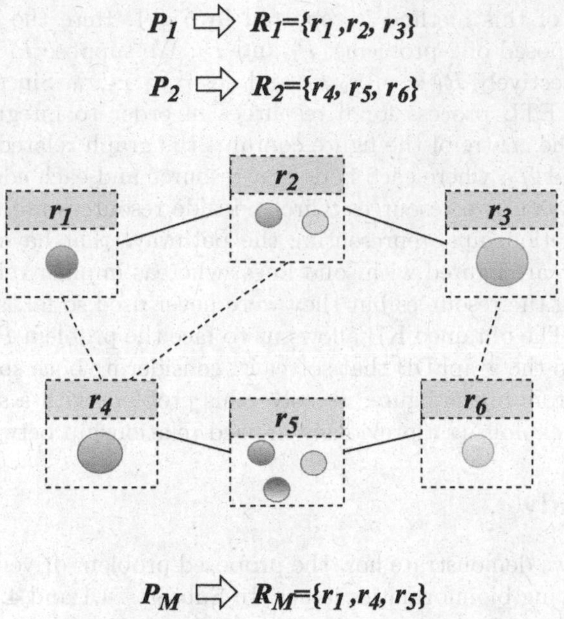

$$P_1 \Rightarrow R_1 = \{r_1, r_2, r_3\}$$

$$P_2 \Rightarrow R_2 = \{r_4, r_5, r_6\}$$

$$P_M \Rightarrow R_M = \{r_1, r_4, r_5\}$$

Fig. 1. An example of the proposed method.

When a minimal set of resources R_1 that solves this problem is given, the second step is to import all the R_1 resources into a graphDB, exploiting the aforementioned ETLs. In this way, we connect all R_1 resources, using their existing relations among each other. The graph representing the problem P_1 can be traversed in order to define a specific path. This path can be obtained by means of a query to the graphDB that solves the first problem. The third step is to analyse the second problem P_2 and to verify if a proper query is able to solve this problem, or in other words, if a pathway in the graphDB that satisfies this query exists. If all the R_1 resources that have been integrated in previous steps are not enough, we need to identify a set of resources R_2 useful for solving P_2. The fourth step is to import all the R_2 resources into the graphDB, just like the second step. However, since R_2 can contain some shared resources with R_1, we perform the ETL process only for $R_2 \backslash R_1$ resources. At this point, we find all available relationships existing among all the $R_1 \cap R_2$ resources, and integrate them into the KB. We iterate the last two steps for N problems $(P_1, P_2, \ldots, P_N)$ in order to populate our graphDB with $R_1 \cap R_2 \cap \ldots R_{N-1} \cap R_N$ resources.

At the end of this training phase, we obtain a graph and, for each problem we have at least a specific pathway that solves it. In addition, we can solve further problems that have not been used to build the knowledge base, traversing the graph in different ways and also exploiting relations (edges) never used before. In other words, we can solve all those problems for which at least a pathway in the graphDB exists.

An example of this method is reported in Fig. 1. Here the training set of problems is composed of 2 problems: P_1 and P_2. We suppose P_1 and P_2 can be solved with respectively $R_1 = r_1, r_2, r_3$ and $R_2 = r_4, r_5, r_6$. Since $R_1 \cap R_2 = 0$, we perform the ETL process for 6 resources in order to integrate them into the graphDB. The centre of the figure contains the graph related to the import process of R_1 and R_2, where each node is a resource and each edge represents a relationship between two resources. Circles inside resources are biological entities. Explicit relationships, representing the pathways that have been used for solving P_1 or P_2, are figured with solid lines, whereas implicit relationships, i.e. they exist among the resources but they were never used so far, are represented as dashed lines. The obtained KB allows us to face the problem P_M, if at least a pathway exists in the graphDB that solves it, considering both solid and dashed lines. In the bottom of the figure, we solve this problem with a set of resources $R_M = r_1, r_4, r_5$, exploiting a previously unused relationship between r_1 and r_4.

4   Case Study

In this Section, we demonstrate how the proposed problem-driven paradigm can be useful for solving bioinformatics issues. In Subsects. 4.1 and 4.2 we introduce two interesting biological case studies that can be solved with different publicly available bioinformatics resources. Then, in Subsect. 4.3 we show how the resources used in previous scenarios are integrated in a graphDB, according to available relationships. Finally, in Subsect. 4.4 we present a problem that can be solved with the existing resources, exploiting a never used connection among two resources.

4.1   Scenario n.1: Target Analysis of Differentially Expressed (DE) MiRNAs in Cancer

miRNAs are small non coding RNAs that are negative regulators of gene expression at post-transcriptional level. They act through selective and partial base-paring mainly of 3'untranslated region (3'UTR) of RNA messenger (mRNA) target [6]. There is a growing evidence of their role as biomarkers in different diseases as cancer [12,21]. For this reason, all the actors of the deregulated event that trigger and guide the development of cancer need to be deeply investigated. We suppose to have a list of DE miRNAs (up/down regulated) linked to a specific disease, e.g. breast cancer. Through specific web resources as miRCancer it is possible to select and extract this miRNA list from the disease of interest. By means of miRbase, it is possible to collect information about those miRNAs, such as their sequences. The next step is the identification of predicted miRNA targets through miRNA-target interaction tools. There are different databases available to this aim as miRanda [3]. This target prediction tool is also linked to NCBI gene, which contains information about genes ID, sequence, gene locus etc.

In Fig. 2 we show how all the analysis steps are linked each other. The type of relationship involved between two biological entities is underlined. From the figure it can be noticed that miRBase has information about both (precursor) miRNA and mature miRNA.

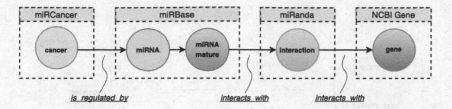

Fig. 2. Used resources and their links of the first scenario.

4.2   Scenario n.2: Analysis of Protein Functions and Pathways

The analysis of protein function is a very important task in biology, especially in the context of disease. Indeed many pathological conditions are consequence of alterations of protein functions. Starting from a specific protein or a group of proteins, from Uniprot services are obtained information strictly connected with the protein like amino acid sequence, structure etc. Resources like Reactome can evidence the cellular context which the protein belongs to. It provides a list of pathways as well as their related molecules and compounds. Finally, by means of the GO functional annotations, it is possible to analyse the molecular functions, cellular components and biological processes of the considered proteins.

In Fig. 3 we show the links about the three resources used in this scenario.

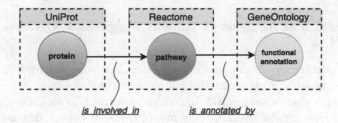

Fig. 3. Used resources and their links of the second scenario.

4.3   Resources Integration

The previous scenarios are related to two different bioinformatics problems. Each one exploits a proper set of available resources and both of them do not use any common resource. At this point, as explained in Sect. 3, we perform an ETL process in order to import all the resources into the proposed graphDB. Result of this process is showed in Fig. 4. The graphical representation includes both the elements and relationships derived from resources used in previous scenarios. According to Figs. 2 and 3, all the used connection are represented with solid lines, whereas those relationships that can be extracted from available resources, but that have not been used yet, are represented with dashed lines. As previously said, a path in this graph represents a possible solution for a specific problem and each line can be traversed in both way.

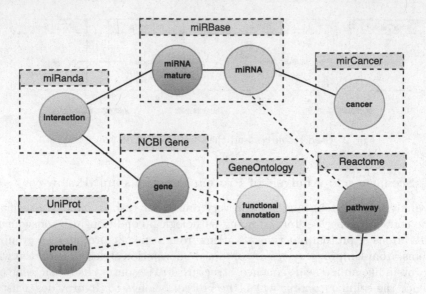

Fig. 4. Graph representing all the resources and their relationships for the proposed first two scenarios. The solid lines are relationship needed for solving the first two scenarios. The dashed lines mean that relationship exists between two resources, but they have not been exploited earlier.

4.4   Scenario n.3: Analysis of Tumour Suppressor/Oncogenic MiRNAs

The integration of all the resources used in scenario n.1 and 2, allows to solve a third biological problem: the analysis of tumour suppressor/oncogenic miRNAs. As previously said, miRNAs can be important biomarkers due to their relevance in gene regulation and their involvement in cancer disease. A group of proteins involved in a specific cellular pathway can be detected using Reactome. This set of proteins is then analysed through miRNA-target interaction tools to identify miRNAs that are predicted targets of those protein products. Together with the use of these resources, Uniprot, HGNC and NCBI gene are also used, giving related information on genes protein products. The list of target can be finally related with a specific disease and investigated by studying the differential expression through resources as miRbase and mirCancer. Indeed, this last web service evidences the relationship between up or down regulated miRNAs in different types of cancer. All the resources and their relationships used in this scenario are shown in Fig. 5. With the dashed arrows we indicate those relationships that have been made available because of the integration of the resources used in the previous scenarios. In this case, there is a possible fork in the implementation of this scenario. If predicted gene targets are considered, then the lower part of the diagram is crossed. On the other hand, the Reactome database provides a list of validated target genes involved in the pathways.

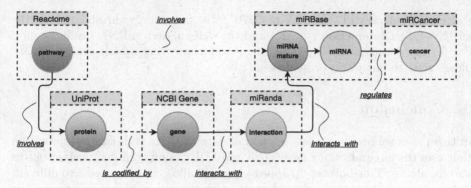

Fig. 5. Used resources and their links of the third scenario. In this case the dashed arrows indicate relationships between resources that have not been considered for solving the two previous scenarios.

5   Graph Query Language

The resources collected and organized in a graphDB, as described in the previous sections, can be queried in order to implement and to get response about the proposed scenarios [11]. One of the most popular query language for graphDB is the Gremlin graph traversal language [2,20]. Gremlin is functional language able to get and edit data organized in a graph structure. Gremlin queries are called graph traversals, because they, in a very compact way, allow to build a path in the graph in order to extract the desired information. In our proposed approach, in which each scenario represents a path over the graph composed of the set of collected resources, Gremlin is a very useful instrument to implement the proposed scenario. For example, according to the scenario n.3 (Sect. 4.4) the related gremlin query is shown in Fig. 6, where we customized the scenario

```
$ ./gremlin.sh

         \,,,/
         (o o)
-----oOOo-(_)-oOOo-----
gremlin> g = new OrientGraph('remote:localhost/biograph')
mag 27, 2016 7:53:55 AM com.orientechnologies.common.log.OLogManager log
INFORMAZIONI: OrientDB auto-config DISKCACHE=13.880MB (heap=455MB os=16.384MB disk=812.209MB)
==>orientgraph[remote:localhost/biograph]
gremlin> pathwayId = 'R-HSA-1640170'
==>R-HSA-1640170
gremlin> freeEnergy = -32
==>-32
gremlin> mirnaProfile = 'up'
==>up
gremlin> g.getVerticesOfClass('pathway')._().has('pathwayId',pathwayId).out('contains').
in('coding').in('interactingGene').filter{it.energy < freeEnergy}.out('interactingMiRNA').
dedup().in('precursorOf').inE('cancer2mirna').filter{it.profile == mirnaProfile}.outV().dedup().
name.order
==>acute myeloid leukemia
==>breast cancer
==>gastric cancer
==>glioma
==>nasopharyngeal carcinoma
==>non-small cell lung cancer
```

Fig. 6. Gremlin code and results for the analysis proposed in scenario n.3

stating the pathway id ("R-HSA-1640170"), the free-energy threshold about the miRNA-target interaction ("−32") and the deregulated miRNA profile ("up"). For further details about Gremlin language for querying a biological graphDB, please refer to our work, presented in [10].

6   Conclusion

In this paper we propose a new method of data analysis and data interpretation that uses the integration, by means of a graphDB, among different web resources and databases. This bottom-up approach would allow an easy access to different independent analysis tools, a rapid exploration through different resources using a common interface and query language, the creation of a knowledge base to explore and solve other problems not evidenced before. Thanks to flexibility of the method, we provide the user not only a methodology able to easily and quickly move thorough very different resources, but also to solve novel biological tasks taking advantage of existing relationships among the integrated resources.

References

1. Amberger, J.S., Bocchini, C.A., Schiettecatte, F., Scott, A.F., Hamosh, A.: Omim. org: online mendelian inheritance in man (omim®), an online catalog of human genes and genetic disorders. Nucleic Acids Res. **43**(D1), D789–D798 (2015)
2. Apache tinkerpop: the gremlin traversal language. https://github.com/tinkerpop/gremlin/wiki
3. Betel, D., Wilson, M., Gabow, A., Marks, D.S., Sander, C.: The microrna. org resource: targets and expression. Nucleic Acids Res. **36**(suppl 1), D149–D153 (2008)
4. Bhattacharya, A., Ziebarth, J.D., Cui, Y.: Polymirts database 3.0: linking polymorphisms in micrornas and their target sites with human diseases and biological pathways. Nucleic Acids Res. **42**(D1), D86–D91 (2014)
5. Botstein, D., Cherry, J.M., Ashburner, M., Ball, C., Blake, J., Butler, H., Davis, A., Dolinski, K., Dwight, S., Eppig, J., et al.: Gene ontology: tool for the unification of biology. Nat. Genet. **25**(1), 25–29 (2000)
6. Brennecke, J., Stark, A., Russell, R.B., Cohen, S.M.: Principles of microrna-target recognition. PLoS Biol. **3**(3), e85 (2005)
7. Croft, D., Mundo, A.F., Haw, R., Milacic, M., Weiser, J., Wu, G., Caudy, M., Garapati, P., Gillespie, M., Kamdar, M.R., et al.: The reactome pathway knowledgebase. Nucleic Acids Res. **42**(D1), D472–D477 (2014)
8. Dennis Jr., G., Sherman, B.T., Hosack, D.A., Yang, J., Gao, W., Lane, H.C., Lempicki, R.A., et al.: David: database for annotation, visualization, and integrated discovery. Genome Biol. **4**(5), P3 (2003)
9. Dweep, H., Gretz, N.: Mirwalk 2.0: a comprehensive atlas of microrna-target interactions. Nat. Methods **12**(8), 697–697 (2015)
10. Rojas, I., Ortuño, F. (eds.): IWBBIO 2017. LNCS, vol. 10209. Springer, Cham (2017)

11. Fiannaca, A., La Paglia, L., La Rosa, M., Messina, A., Storniolo, P., Urso, A.: Integrated DB for bioinformatics: a case study on analysis of functional effect of MiRNA SNPs in cancer. In: Renda, M.E., Bursa, M., Holzinger, A., Khuri, S. (eds.) ITBAM 2016. LNCS, vol. 9832, pp. 214–222. Springer, Cham (2016). doi:10.1007/978-3-319-43949-5_17

12. Fiannaca, A., La Rosa, M., La Paglia, L., Rizzo, R., Urso, A.: Analysis of mirna expression profiles in breast cancer using biclustering. BMC Bioinform. 16(Suppl 4), S7 (2015)

13. Gray, K.A., Yates, B., Seal, R.L., Wright, M.W., Bruford, E.A.: Genenames.org: the HGNC resources in 2015. Nucleic Acids Res. 43(D1), D1079–D1085 (2015)

14. Huang, D.W., Sherman, B.T., Lempicki, R.A.: Bioinformatics enrichment tools: paths toward the comprehensive functional analysis of large gene lists. Nucleic Acids Res. 37(1), 1–13 (2009)

15. Kanehisa, M., Sato, Y., Kawashima, M., Furumichi, M., Tanabe, M.: Kegg as a reference resource for gene and protein annotation. Nucleic Acids Res. 44(D1), D457–D462 (2016)

16. Kozomara, A., Griffiths-Jones, S.: MiRBase: integrating microRNA annotation and deep-sequencing data. Nucleic Acids Res. 39(Database issue), D152–7 (2011)

17. Kumar, S., Dudley, J.: Bioinformatics software for biologists in the genomics era. Bioinformatics 23(14), 1713–1717 (2007)

18. Laganà, A., Forte, S., Giudice, A., Arena, M., Puglisi, P.L., Giugno, R., Pulvirenti, A., Shasha, D., Ferro, A.: Miro: a mirna knowledge base. Database 2009, bap. 008 (2009)

19. Mi, H., Poudel, S., Muruganujan, A., Casagrande, J.T., Thomas, P.D.: Panther version 10: expanded protein families and functions, and analysis tools. Nucleic Acids Res. 44(D1), D336–D342 (2016)

20. Rodriguez, M.A.: The gremlin graph traversal machine and language (invited talk). In: Proceedings of the 15th Symposium on Database Programming Languages - DBPL 2015, pp. 1–10. ACM Press, New York (2015)

21. Romero-Cordoba, S.L., Salido-Guadarrama, I., Rodriguez-Dorantes, M., Hidalgo-Miranda, A.: Mirna biogenesis: biological impact in the development of cancer. Cancer Biol. Ther. 15(11), 1444–1455 (2014)

22. Rosenbloom, K.R., Armstrong, J., Barber, G.P., Casper, J., Clawson, H., Diekhans, M., Dreszer, T.R., Fujita, P.A., Guruvadoo, L., Haeussler, M., et al.: The UCSC genome browser database: 2015 update. Nucleic Acids Res. 43(D1), D670–D681 (2015)

23. Schuler, G.D., Epstein, J.A., Ohkawa, H., Kans, J.A.: Entrez: molecular biology database and retrieval system. Methods Enzymol. 266, 141–162 (1996)

24. Sigrist, C.J.A., de Castro, E., Cerutti, L., Cuche, B.A., Hulo, N., Bridge, A., Bougueleret, L., Xenarios, I.: New and continuing developments at PROSITE. Nucleic Acids Res. 41(Database issue), D344–7 (2013)

25. Szklarczyk, D., Franceschini, A., Wyder, S., Forslund, K., Heller, D., Huerta-Cepas, J., Simonovic, M., Roth, A., Santos, A., Tsafou, K.P., Kuhn, M., Bork, P., Jensen, L.J., von Mering, C.: STRING v10: protein-protein interaction networks, integrated over the tree of life. Nucleic Acids Res. 43(Database issue), D447–52 (2015)

26. The Gene ontology consortium: going forward. Nucleic Acids Res. 43(D1), D1049–D1056 (2015)

27. The UniProt Consortium: UniProt: a hub for protein information. Nucleic Acids Res. **43**(D1), D204–D212 (2015)
28. Xie, B., Ding, Q., Han, H., Wu, D.: MiRCancer: a microRNA-cancer association database constructed by text mining on literature. Bioinformatics **29**(5), 638–644 (2013)

Parameter Inference in Differential Equation Models of Biopathways Using Time Warped Gradient Matching

Mu Niu[1]($\boxtimes$), Simon Rogers[3], Maurizio Filippone[4], and Dirk Husmeier[2]

[1] School of Computing, Electronics and Mathematics, Plymouth University, Plymouth, UK
mu.niu@plymouth.ac.uk
[2] School of Mathematics and Statistics, University of Glasgow, Glasgow, UK
[3] Department of Computer Science, University of Glasgow, Glasgow, UK
[4] Department of Data Science, Eurecom, France

Abstract. Parameter inference in mechanistic models of biopathways based on systems of coupled differential equations is a topical yet computationally challenging problem due to the fact that each parameter adaptation involves a numerical integration of the differential equations. Techniques based on gradient matching, which aim to minimize the discrepancy between the slope of a data interpolant and the derivatives predicted from the differential equations, offer a computationally appealing shortcut to the inference problem. Gradient matching critically hinges on the smoothing scheme for function interpolation, with spurious oscillations in the interpolant having a dramatic effect on the subsequent inference. The present article demonstrates that a time warping approach that aims to homogenize intrinsic functional length scales can lead to a significant improvement in parameter estimation accuracy. We demonstrate the effectiveness of this scheme on noisy data from a dynamical system with periodic limit cycle, and a biopathway model.

Keywords: Biopathways · Differential equations · Gradient matching · Reproducing kernel hilbert space · Time warping · Optimisation

1 Scientific Background

The elucidation of the structure and dynamics of biopathways is a central objective of systems biology. A standard approach is to view the biopathway as a network of biochemical reactions, modelled as a system of ordinary differential equations (ODEs). This system can typically be expressed as:

$$\dot{x} = \frac{dx}{dt} = f\left(x(t), \theta\right), \tag{1}$$

where $x = (x_1, \ldots, x_r)$ is a time-dependent vector of r state variables, and the parameters θ determine system kinetics. For complex biopathways, only a small

© Springer International Publishing AG 2017
A. Bracciali et al. (Eds.): CIBB 2016, LNBI 10477, pp. 145–159, 2017.
DOI: 10.1007/978-3-319-67834-4_12

fraction of the parameters θ can typically be measured and the major proportion of kinetic parameters has to be inferred from observed (typically noisy and sparse) time course concentration profiles. In principle, this can be accomplished with standard techniques from machine learning and statistical inference. These techniques are based on quantifying the difference between predicted and measured time course profiles by some appropriate metric, to obtain the likelihood of the data. The parameters are then optimised to maximize the likelihood (or a regularised version thereof). However, the nature of the ODE-based model (1) renders the inference problem computationally challenging in two respects. Firstly, for nonlinear functions $f(.)$, the ODE system (1) usually does not permit a closed-form solution. One therefore has to resort to numerical integration every time the kinetic parameters θ are adapted, which is computationally onerous. Secondly, the likelihood function in the space of parameters θ is typically *not* unimodal, but suffers from multiple local optima. Hence, even if a closed-form solution of the ODEs existed, inference by maximum likelihood would be NP-hard, calling for a computationally expensive iterative optimisation.

To circumvent the excessive computational complexity of explicitly solving the ODE system, as described above, various authors have adopted an approach based on gradient matching (Ramsay et al. 2007, Xun et al. 2013, Calderhead et al. 2009, Dondelinger et al. 2013, Macdonald et al. 2015, González et al. 2013, 2014). Gradient matching is based on the following two-step procedure. In a first *smoothing* step, obtain an estimator of the solution directly from the data. In a second *inference* step, estimate the kinetic parameters θ by optimising a functional criteria constructed from the difference between the slope from the estimated solution and the θ-dependent time derivative from the ODEs. With gradient matching, the ODEs never have to be solved explicitly, and the initial conditions do not have to be inferred. However, a problem intrinsic to this approach is the critical dependence of the inference scheme on the form of the initial interpolant. Small 'wiggles', which are hardly discernible at the level of the interpolant itself, can have dramatic effects at the level of the derivatives, which determine the parameter estimation. For noisy data, an adequate smoothing scheme is essential. Any smoothing scheme is based on intrinsic functional length scales, though, and these length scales may vary in time.

Here we present a new method that aims to improve the initial interpolants through homogenizing the intrinsic length scales. The key idea is that a regular sinusoid is easy to learn, whereas a quasi-periodic signal with varying frequencies is not. The objective, hence, is to find a warping of the time axis that counteracts the inhomogeneity in the period. This can easily be effected in principle. The characteristic feature of a regular sinusoid is the proportionality of the original function to its second derivative. Hence, we need to find a bijective transformation of time such that some metric quantifying the difference between the original function and a rescaled version of its second derivative is minimized in warped time. The procedure thus reduces to a double minimization problem, with respect to both the parameters of the map and the scaling parameter. Although time warping has been used in speech recognition for constructing robust speech recognisers undeterred by the variability in pitch (Sakoe and Chiba, 1978) and in systems biology to automatically recognize and align important genomic features (Lukauskas et al. 2016), to the best of our knowledge this is the first time it has been proposed to improve inference in ODE systems.

2   Materials and Methods

Consider a dynamical system consisting of r interacting states x_s, $1 \leq \overset{\circ}{s} \leq r$. For example, in our application, the state variables represent concentrations of protein isoforms (in the Biopathway model) and membrane potentials (in the FitzHugh-Nagumo model). We assume that we have time series of n noisy observations $\boldsymbol{y}_s = (y_{s1}, \ldots, y_{sn})'$ of the states $\boldsymbol{x}_s = (x_{s1}, \ldots, x_{sn})'$, subject to iid additive Gaussian noise $\boldsymbol{\epsilon}_s \sim N(0, \sigma^2 \boldsymbol{I})$:

$$\boldsymbol{y}_s = \boldsymbol{x}_s + \boldsymbol{\epsilon}_s. \tag{2}$$

The objective of inference is to learn $\boldsymbol{\theta}$ from these noisy measurements. We adopt an approach based on reproducing kernel Hilbert spaces (RKHS), where functions are expressed as a linear combination of kernel functions evaluated at the data points

$$x(t) = \sum_{i=1}^{n} b_i k(t, t_i), \tag{3}$$

where $b_i \in \mathbb{R}$ and t_i is the ith time point. The sth component of the dynamical system at time t can be modelled as

$$g_s(t; \boldsymbol{b}_s) = \sum_{i=1}^{n} b_{si} k(t, t_i), \tag{4}$$

with derivatives

$$\dot{g}_s(t; \boldsymbol{b}_s) = \sum_{i=1}^{n} b_{si} \frac{\partial k(t, t_i)}{\partial t} = \sum_{i=1}^{n} b_{si} \dot{k}(t, t_i) \tag{5}$$

$$\ddot{g}_s(t; \boldsymbol{b}_s) = \sum_{i=1}^{n} b_{si} \frac{\partial^2 k(t, t_i)}{\partial t^2} = \sum_{i=1}^{n} b_{si} \ddot{k}(t, t_i). \tag{6}$$

$\dot{\boldsymbol{g}}(t_i)$ is the vector form of gradient estimates for all ODEs states at time t_i. The ODE parameter $\boldsymbol{\theta}$ can then be estimated by minimizing the difference between $\dot{\boldsymbol{g}}(t_i)$ and the gradient predicted from the ODEs, $\boldsymbol{f}(\boldsymbol{g}(t_i), \boldsymbol{\theta})$, using the following loss function:

$$L(\boldsymbol{\theta}) = \sum_{s=1}^{r} \sum_{i=1}^{n} \left[\dot{g}_s(t_i) - f_s(g(t_i), \boldsymbol{\theta}) \right]^2 \tag{7}$$

In order to overcome the difficulties caused by variations in the intrinsic functional length scales on the interpolation, we introduce a two-layer approach. The objective of the first layer is to transform, for each of the variables s of the dynamical system, time t via a bijection $\tilde{t} = w_s(t)$[1] such that in warped time $\tilde{t}$, the unknown solutions x_s of the dynamical system show less variation in their intrinsic length scales. More specifically, we aim to transform the target function into a regular sinusoid by exploiting the fact that a sinusoid is closed under second-order differentiation (subject to a rescaling). We define the transformation of time as

$$\tilde{t} = w_s(t, \boldsymbol{b}^w, l^w) = \sum_{j=1}^{n} \exp\left(b_j^w\right) \mathcal{S}(t - t_j, l^w); \qquad \mathcal{S}(z, l^w) = \frac{1}{1 + \exp(-l^w z)}, \quad (8)$$

where the strict monotonicity of $\mathcal{S}(.)$ and the non-negativity of $\exp(.)$ guarantee bijectivity. The superscript w indicate the kernel parameter of the basis function and coefficients for the warping function. The number of basis functions n can, in principle, be treated as a model selection problem. In practice, we found that setting n to the actual number of observations gave satisfactory results (as reported in Sect. 3). In the original time domain, the sth variable of the dynamical system, $x_s(t)$, is approximated by the smooth interpolant $g_s(t)$. This function is now transformed, by virtue of the bijection (8), into $q_s(\tilde{t})$, where

$$g_s(t) = q_s \circ w_s(t) = q_s(\tilde{t}) \tag{9}$$

and $w_s(t)$ is shorthand notation for the bijection defined in (8).

This results in a four step scheme for ODE parameter estimation:

Step 1: Initialization. We initialize the system with standard kernel ridge regression. This gives us smooth interpolants of the observed states $g_s(t)$ in the original time domain t. We then initialize $\tilde{t} = t$ and $g_s(t) = q_s(\tilde{t})$, for each of the variables s of the dynamical system in turn.

Step 2: Time warping. The bijection between the original time domain $t \in [T_0, T_1]$ and the warped domain $\tilde{t} \in [\tilde{T}_0, \tilde{T}_1]$ is obtained by minimising the objective function

$$L_w = \int \left(\ddot{q}_s(\tilde{t}) + [\lambda^w]^2 q_s(\tilde{t})\right)^2 d\tilde{t} + \lambda_t \left(\left(\tilde{T}_1 - T_1\right)^2 + \left(\tilde{T}_0 - T_0\right)^2\right). \tag{10}$$

The first term is minimized if $q_s(\tilde{t})$ is a regular oscillation (i.e. phase-shifted cosine or sinusoid) with angular frequency λ^w. In practice, we will often have prior knowledge about typical periods of oscillation which can easily be incorporated by restricting the domain of λ^w, e.g. by modelling it as the output of a rescaled sigmoidal function. The second term is a regularisation term, weighted by a penalty parameter $\lambda_t > 0$, to discourage degenerate solutions. The practical choice of λ_t is not critical as long as it is sufficiently large (we increase λ_t until the results are invariant wrt a further increase).

---

[1] Recall that $\tilde{t}$ depends on s, so a more accurate (but cumbersome) notation would be $\tilde{t} \to \tilde{t}_s$.

The integral in (10) is analytically intractable and needs to be solved numerically:

$$L_w = \sum_{i=2}^{n} \left(\ddot{q}_s(\tilde{t}_i) + [\lambda^w]^2 q_s(\tilde{t}_i) \right)^2 \Delta t_i + \lambda_t \left(\left(\tilde{T}_1 - T_1 \right)^2 + \left(\tilde{T}_0 - T_0 \right)^2 \right) \quad (11)$$

where $\Delta t_i = t_i - t_{i-1}$ and the parameters λ^w, l^w and b^w are optimised iteratively until some convergence criterion is met.

Step 3: Interpolation. The second layer deals with function interpolation. The original data points $y_s(t_i)$ are mapped to the warped time points, $y(\tilde{t}_i)$. We then apply standard kernel ridge regression with an RBF kernel in the warped domain, which gives us the smooth interpolant $q_s(\tilde{t})$, for each of the variables s in the dynamical system in turn:

$$q_s(\tilde{t}; \boldsymbol{b}_s^q) = \sum_{j=1}^{n} b_{sj}^q k(\tilde{t}, \tilde{t}_j). \quad (12)$$

Note that this interpolation problem is less susceptible to overfitting or oversmoothing, due to the fact that the intrinsic functional length scales (i.e. periods for an oscillating signal) have been homogenized by virtue of the time warping. Unwarping $q_s(\tilde{t})$ back into the original time domain t is straightforward since $w_s(t)$ is bijective. We have $g_s(t) = q_s(\tilde{t})$, and

$$\frac{dg_s(t)}{dt} = \frac{dq_s(\tilde{t})}{dt} = \sum_{j=1}^{n} b_{sj}^q \frac{\partial k(\tilde{t}, \tilde{t}_j)}{\partial \tilde{t}} \frac{d\tilde{t}}{dt} = \sum_{j=1}^{n} b_{sj}^q \frac{\partial k(\tilde{t}, \tilde{t}_j)}{\partial \tilde{t}} w_s'(t). \quad (13)$$

Step 4: Gradient matching. Finally, we estimate the ODE parameters with gradient matching, i.e. by minimizing the following objective function with respect to $\boldsymbol{\theta}$:

$$L(\boldsymbol{\theta}) = \sum_{s=1}^{r} \sum_{i=1}^{n} \left[\dot{g}_s(t_i) - f_s(\boldsymbol{g}(t_i), \boldsymbol{\theta}) \right]^2 = \sum_{s=1}^{r} \sum_{i=1}^{n} \left[\frac{dq_s(\tilde{t}_i)}{d\tilde{t}_i} \frac{d\tilde{t}_i}{dt_i} - f_s(\boldsymbol{q}(\tilde{t}_i), \boldsymbol{\theta}) \right]^2 \quad (14)$$

3   Results

The objective of our study is to evaluate the performance improvement of the novel two-level time-warping method proposed in Sect. 2 over the standard RKHS gradient matching method summarized in Sect. 1. This method is akin to the one proposed in (González et al. 2013, 2014) and hence representative of the current state of the art. For notational convenience, we refer to these methods as RKGW (W for warping) and RKG, respectively and compare them on data generated from two different ODE systems. The ODEs were numerically integrated using a low-order Runge-Kutta method with automatic step-size adjustment, using the R function ODE23s. The timepoints produced by ODE23s were then uniformly downsampled by 50%, keeping every 2nd output from ODE23s (leading to $n = 37$ observations per state for the FitzHugh-Nagumo model and $n = 17$ for the biopathway model). The true state values from each model were repeatedly and independently subjected to additive iid Gaussian noise from 50 independent noise instantiations over a range of signal-to-noise ratios (SNR). Each of these 50 data realisation is used for parameter inference, and the results are collated.

FitzHugh-Nagumo. The FitzHugh-Nagumo system is a two-dimensional dynamical system used for modelling spike generation in axons (FitzHugh, 1955). It has two state variables, x_1 and x_2, and three parameters: a, b and c. We numerically solved the ODEs for $a = 0.2$, $b = 0.2$, $c = 3$, $t \in (0, 10)$, and initial conditions $x_1(0) = 0.5$ and $x_2(0) = 1$. As already mentioned, numerical solving and down-sampling resulted in states being observed at $n = 37$ distinct time points.

$$\dot{x}_1 = c \cdot \left(x_1 - x_1^3/3 + x_2\right), \qquad \dot{x}_2 = -c^{-1}\left(x_1 - a + b \cdot x_2\right) \qquad (15)$$

Biopathways. The biopathway model describes the interaction of five protein isoforms, S, dS, R, RS, Rpp, in a signal transduction pathway and was previously studied by Vyshemirsky and Girolami [2008]. Changes in protein abundance over time is described by a combination of mass action and Michaelis-Menten kinetics:

$$[\dot{S}] = -k_1 \cdot [S] - k_2 \cdot [S] \cdot [R] + k_3 \cdot [RS]$$
$$[\dot{dS}] = k_1 \cdot [S]$$
$$[\dot{R}] = -k_2 \cdot [S] \cdot [R] + k_3 \cdot [RS] + \frac{k_5 \cdot [Rpp]}{k_6 + [Rpp]}$$
$$[\dot{RS}] = k_2 \cdot [S] \cdot [R] - k_3 \cdot [RS] - k_4 \cdot [RS]$$
$$[\dot{Rpp}] = k_4 \cdot [RS] - \frac{k_5 \cdot [Rpp]}{k_6 + [Rpp]} \qquad (16)$$

The square brackets, $[\cdot]$, denote concentrations of the protein isoforms (the states), and $k_{1:6}$ represent the 6 kinetic parameters to be inferred. It turns out that k_5 and k_6 are only weakly identifiable, and we have thus assessed the accuracy of inference based on the ratio $\frac{k_5}{k_6}$. We numerically solved the ODEs for $k_1 = 0.07$, $k_2 = 0.6$, $k_3 = 0.05$, $k_4 = 0.3$, $k_5 = 0.017$, $k_6 = 0.3$, $t \in (0, 100)$, with initial conditions $S(0) = 1$, $dS(0) = 0$, $R(0) = 1$, $RS(0) = 0$ and $Rpp(0) = 0$. This generated $n = 17$ data points.

The true solutions of these two systems are shown in Fig. 1.

Figure 2 shows a graphical demonstration of the warping process using the state S of the Biopathway system as an example. RBF regression (Fig. 2(b)) is unable to cope with both the rapid drop and the saturated section. In Fig. 2(c) we show the interpolation in the warped time domain and in (d) the warped interpolant in the original space. The improvement over the interpolant shown in Fig. 2(b) is clear.

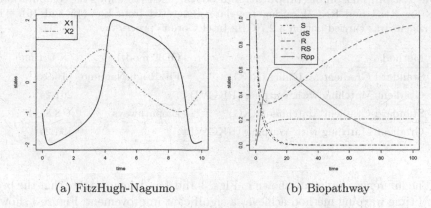

(a) FitzHugh-Nagumo   (b) Biopathway

Fig. 1. True solutions of the ODE systems. Note the temporal inhomogeneity of the intrinsic length scales.

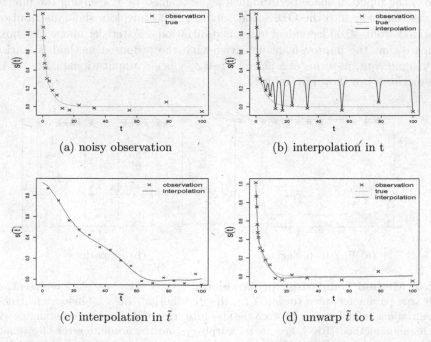

(a) noisy observation   (b) interpolation in t

(c) interpolation in $\tilde{t}$ (d) unwarp $\tilde{t}$ to t

Fig. 2. Warping example. (a) The true signal and 10db SNR noisy data. (b) The initial interpolation using RBF kernel RKHS regression. Due to the non-stationary length scale of the signal, the RBF is unable to produce a sensible interpolation. (c) The interpolation in the warped time domain using an RBF kernel. (d) The interpolation result from (b) unwarped and in the original space. Note the clear improvement over the interpolant in (b).

Table 1. Comparison of computational costs. The table shows the computational costs for RKG and RKGW, using the data generated from Eqs. (15 and 16). The experiment was carried out on a 2.7 GHz Intel Core i5 processor.

Method	ODE model	CPU time
Standard Gradient Matching (RKG)	FitzHugh-Nagumo	16.9s
Gradient Matching with Warping (RKGW)		261.2s
Standard Gradient Matching (RKG)	Biopathways	8.8s
Gradient Matching with Warping (RKGW)		194.2s

The inference results are shown in Figs. 3 and 4 and demonstrate that the proposed time warping method achieves a significant improvement. Figure 3 shows, for each parameter and each benchmark system, the difference of the absolute differences between the inferred and the true parameters. The difference between the methods is defined such that positive values indicate that the proposed warping method (RKGW) outperforms the reference method (RKG) (as explained above, the function space performance was obtained by reinserting the inferred parameters back into the ODEs and solving). The boxplots show distributions obtained from 50 independent data instantiations. Asterisks above the boxes indicate that the improvement achieved with the proposed method is statistically significant, in terms of a paired t-test. Typical computational costs for the two methods are shown in Table 1.

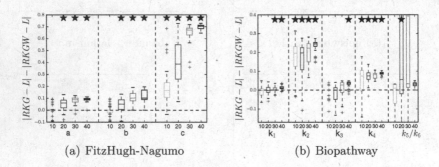

(a) FitzHugh-Nagumo   (b) Biopathway

Fig. 3. Method comparison in parameter space. The box plots represent, for each true parameter value (denoted L), the distribution (from 50 independent noise instantiations) of differences between the absolute error of the parameter estimates with the baseline method (RKG, Sect. 1, no warping), and the absolute error of estimates with the proposed method (RKGW, Sect. 2, with time warping). Positive values (above the dashed horizontal line) indicate that time warping improves performance. The horizontal axis shows different signal-to-noise ratios for each parameter. Asterisks above a box indicate where the performance improvement is significant (based on a paired t-test).

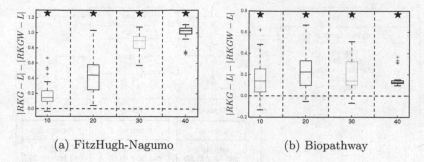

(a) FitzHugh-Nagumo   (b) Biopathway

Fig. 4. Method comparison in function space. Similar boxplot representation as in Fig. 3, but showing the distribution of the differences between the absolute errors of the function estimates; these function estimates are obtained by inserting the estimated parameters into the ODEs and numerically solving. Positive values indicate that the proposed method outperforms the standard method, asterisks indicate that the improvement is significant (paired t-test).

4   R Package

To maximise utility to the community, we have implemented our warping and gradient matching schemes as a flexible object oriented R package. This allows for easy re-use and reproducibility and the package is described below. The package is implemented using the object oriented paradigm using R6 classes (Chang, 2016). A UML class diagram of the package is shown in Fig. 5.

Kernel Class: The `Kernel` class represents kernel functions. We provide three kernel options although the modular nature of the code makes it easy to add more. The standard RBF kernel and MLP kernel are implemented in the child classes `RBF` and `MLP`. Although the sigmoid function does not yield a reproducing kernel and is not used for kernel ridge regression, it includes many of the properties of standard kernel functions and hence its inclusion as a subclass of `Kernel`. The sigmoid function is used as the basis function in the `Warp` class.

RKHS Class: The standard kernel ridge regression in step 1 of Sect. 2 is implemented as the `RKHS` class. The `RKHS` class requires an instance of the `Kernel` class. The kernel and weighting parameters of the l^2 norm of RKHS can be estimated using cross-validation, implemented in the `skcross()` operation. Interpolants and gradient of interpolants can be estimated using Eq. 5 which is implemented as the `predict()` operation.

Ode Class: Our code can work with any system of user-specified ODEs, which are stored in the `Ode` class. The user needs to provide the ODEs, the initial condition if they want to numerically solve the ODEs and initial values of the ODE parameters. The gradient estimates from the ODEs themselves are provided by

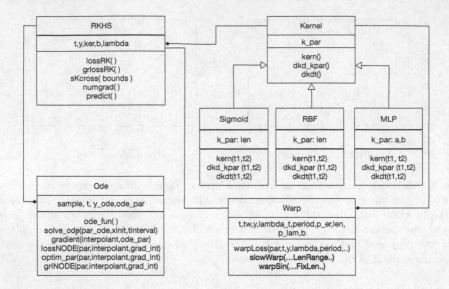

Fig. 5. R6 UML Class diagram. Each block represents a class object. The top compartment is the class name. The middle compartment lists the class's attributes. The bottom compartment lists the class's operations. Inheritance is indicated by a solid line with a closed, unfilled arrowhead pointing at the super class. The solid line with filled diamond arrowhead indicates a composition relationship between two classes. The solid line without arrowhead indicates an association relationship between two classes.

the operation `gradient()`, which takes interpolants and ODE parameters as inputs. By matching gradient estimates of the interpolants and the true ODE, the ODE parameters are estimated using the operation `optim_par()`. We also provide the operation `solve_ode()` to numerically solve ODEs in the `Ode` class. Inference of ODE parameters using standard gradient matching (RKG) requires the `Ode` class and the `Kernel` class only.

Warp Class: The warping scheme in step 2 of Sect. 2 is implemented in the `Warp` class. It takes an instance of the `RKHS` class as an attribute. Interpolants from kernel ridge regression using the `RKHS` class are warped into a sinusoidal signal by minimising the loss function in Eq. 11, which is programmed as the operation `warpLoss()`. The warped time index vector is optimised using the operation `warpSin()`. The improved interpolant can be re-learned with the warped time index using standard kernel ridge regression. The gradient matching can also be calculated with the warped interpolants using the `Ode` class.

4.1   Example

In this section, we provide an example of the code use with the FitzHugh Nagumo system. The FitzHugh Nagumo ODEs are defined as a function `FN_fun`. `x[1]` and `x[2]` are the two states and `a,b,c` the three ODEs parameters in Eq. 15:

```
FN_fun = function(t,x,par_ode){
    a=par_ode[1]
    b=par_ode[2]
    c=par_ode[3]
    as.matrix( c( c*(x[1]-x[1]^3/3 + x[2]),-1/c*(x[1]-a+b*x[2]) ) )
}
```

The FN_fun function is used to generate the testing data by numerically solving the ODEs. A Ode class object is created as FN. The initial condition and the time interval are defined as variables xinit and tinterv. The testing data y_no is generated by adding Gaussian noise to the true ODE solutions y_ode. n_o is the length of the data vector. If the user wants to use a real dataset, this step can be skipped.

```
FN = Ode$new(2,fun=FN_fun)
xinit = as.matrix(c(0.5,1))
tinterv = c(0,6)
FN$solve_ode(c(1,1,4,1),xinit,tinterv)
n_o = length(FN$t)
y_no =  t(FN$y_ode) + rmvnorm(n_o,c(0,0),0.1*diag(2))
```

At first we use the standard gradient matching (RKG) method to estimate the ODE parameters according to step 1 in Sect. 2. The result of the RKG scheme can be used as the initial value for the warping scheme. An RBF class instance is created as ker1. The argument of RBF$new(1) defines the initial length scale parameter, which in this example is set to 1. The kernel hyper-parameters are optimised using cross validation with the RKHS class operation skcross(.). Arguments of rkhs$new(.) define initial values of the functions and the initial value of the l^2 norm weighting parameter. The initial interpolation and gradient of interpolants are calculated using kernel ridge regression with the RKHS class operation predict().

```
ker1 = RBF$new(1)
rk1 = rkhs$new(t(y_no)[1,],FN$t,rep(1,n_o),1,ker1)
rk1$skcross()
pre1 = rk1$predict()
```

An identical procedure is applied for the second state:

```
ker2 = RBF$new(1)
rk2 = rkhs$new(t(y_no)[2,],FN$t,rep(1,n_o),1,ker2)
rk2$skcross()
pre2 = rk2$predict()
```

The interpolation and the gradient estimates of both states are used to estimate the ODE parameters using gradient matching, which is implemented as the Ode class operation optim_par(.). The initial values of the ODE parameters, interpolants and gradient estimates for both states are provided as the arguments of optim_par(.).

```
intp= rbind(pre1$pred,pre2$pred)
grad= rbind(pre1$grad,pre2$grad)
ode_par = FN$optim_par( c(0.1,0.1,0.1,0.1), intp, grad )
```

The warping method introduced in step 2 of Sect. 2 is implemented as follows. p1 is the predefined period of the first state of the ODEs. lambda_t is defined as in Eq. 11. wker1 is defined as an instance of the Sigmoid class. The warping function is modelled as a linear combination of sigmoid basis functions by defining wp1 as an instance of the Warp class. The kernel hyper-parameters and coefficients of each basis function are optimised using the Warp class operation warpSin(.). The argument of warpSin(.) defines the initial value of the lengthscale kernel parameter, which is ini_len1 in this example. The time indices of the warped signal are stored in tw1.

```
p1=6;  eps= 1;  lambda_t= 50
wker1 = Sigmoid$new(1)
wp1 = Warp$new( pre1$pred, FN$t, rep(1,n_o), lambda_t, wker1)
ini_len1 = 3
tw1 = wp1$warpSin(ini_len1, p1, eps)
```

An identical procedure is applied to the second state:

```
p2=5.5;
wker2 = Sigmoid$new(1)
wp2 = Warp$new( pre2$pred, FN$t, rep(1,n_o), lambda_t, wker2)
ini_len2 = 3
tw2 = wp1$warpSin(ini_len2, p2, eps)
```

The warping function is monotonically increasing and can be learned with a linear combination of sigmoid functions. However, the estimation of the gradient of the warping function using sigmoid functions may be poor as the interpolation from the sigmoid basis function is not smooth. To overcome this, we re-learn the gradient of the warping function using standard kernel ridge regression with an MLP kernel. A MLP class instance is created as mker1. The kernel hyper-parameters are optimised using cross-validation with the RKHS class operation skcross(.). Arguments of skcross(.) define the upper and lower bound of kernel hyperparameters:

```
mker1 = MLP$new(c(1,1))
rkm1 = rkhs$new(t(tw1$tw),FN$t,rep(1,n_o),1,mker1)
rkm1$skcross(c(0.001,1000))
tw1 = rkm2$predict()

mker2 = MLP$new(c(1,1))
rkm2 = rkhs$new(t(tw2$tw),FN$t,rep(1,n_o),1,mker2)
rkm2$skcross(c(0.001,1000))
tw2 = rkm2$predict()
```

In step 3 of Sect. 2, the improved interpolants are learned in the warped time domain. The code for this step is similar to step 1, however the time indexes are now changed to the warped time indices `tw1$pred`.

```
ker1 = RBF$new(1)
rk1 = rkhs$new(t(y_no)[1,],tw1$pred ,rep(1,n_o),1,ker1)
rk1$skcross()
pre1 = rk1$predict()

ker2 = RBF$new(1)
rk2 = rkhs$new(t(y_no)[2,],tw2$pred,rep(1,n_o),1,ker2)
rk2$skcross()
pre2 = rk2$predict()
```

Finally, both the improved interpolants and the gradient estimates are used for gradient matching. By applying the chain rule in Eq. 14, the gradient of the interpolant in the original time domain becomes the product of the gradient in the warped time domain and the gradient of the warping function `pre1$grad*tw1$grad`. The ODE parameters can be learned by using the `Ode` class operation `optim_par(.)`

```
intp= rbind(pre1$pred,pre2$pred)
grad= rbind(pre1$grad*tw1$grad,pre2$grad*tw2$grad)
FN$optim_par( c(0.1,0.1,0.1,0.1), intp, grad)
```

The parameter estimates generated from this example can be used to produce the boxplot in Fig. 3(a).

5   Conclusion

Carrying out parameter inference in models described by ODEs is a challenging problem, due to the need to repeatedly perform computationally expensive numerical integration to solve the ODEs. While gradient matching approaches mitigate this issue, their success critically hinges on the quality of the initial interpolation scheme. In cases where the solutions to the ODE systems exhibit nonstationarity and substantial variations of intrinsic length scales, standard RKHS or Gaussian process approaches typically fail to accurately represent the unknown true functions, leading to poor ODE parameter estimates. In this paper, we have proposed a remedy for this problem by combining gradient matching techniques and time warping. The latter, in particular, is inspired by the work in Calandra et al. (2016), where Gaussian processes are made nonstationary by a reparameterisation of the input space. In our work, we use an RKHS interpolation approach instead, and learn the reparameterisation by optimising a separate objective function that aims to homogenize the intrinsic functional length scales. We have demonstrated that the proposed time warping is effective in improving the quality of gradient matching approaches in two applications representative of biological dynamical systems, one with a limit cycle, the other with a stable equilibrium point.

Our work proposes a first proof of concept that time warping can be useful to improve parameter inference in ODE models. We are currently investigating extensions of our work in the direction of including some form of regularisation in the estimation of the parameters based on the structure of the ODEs. This, for example, could come in the form of alternating between revising the interpolant in light of the estimated ODE parameters and the estimation of the ODE parameters, or in the form of a prior, following, e.g., the work on hierarchical Bayesian models in Xun et al. (2013).

Our current work has focused on fully observed data. If state variables are unobserved, then they have to be treated as latent variables, for which standard inference procedures are available. See, for instance, Sect. 5.3 in (Calderhead et al. 2009), and Sect. 4 in (Macdonald et al. 2015).

To facilitate code re-use and reproducability, we have provided a flexible implementation of the system to allow others to use our algorithm, reproduce our results, or use our algorithm to benchmark newer approaches. The code has been built in a modular, object oriented manner allowing flexibility and optimising the opportunities for code re-use. The R package described in this paper is available at `http://dx.doi.org/10.5525/gla.researchdata.383`.

Acknowledgments. This work was supported by EPSRC (EP/L020319/1). MF gratefully acknowledges support from the AXA Research Fund.

References

Calandra, R., Peters, J., Rasmussen, C.E., Deisenroth, M.P.: Manifold gaussian processes for regression. ArXiv e-prints, February 2016

Calderhead, B., Girolami, M., Lawrence, N.D.: Accelerating bayesian inference over nonlinear differential equations with gaussian processes. In: Advances in Neural Information Processing Systems, pp. 217–224 (2009)

Chang, W.: R6: classes with reference semantics, R package version 2.2.0 (2016). https://CRAN.R-project.org/package=R6

Dondelinger, F., Filippone, M., Rogers, S., Husmeier, D.: ODE parameter inference using adaptive gradient matching with Gaussian processes. In: Sixteenth International Conference on Artificial Intelligence and Statistics (2013)

FitzHugh, R.: Mathematical models of threshold phenomena in the nerve membrane. Bull. Math. Biophys. **17**(4), 257–278 (1955)

González, J., Vujačić, I., Wit, E.: Inferring latent gene regulatory network kinetics. Stat. Appl. Genet. Mol. Biol. **12**(1), 109–127 (2013)

González, J., Vujačić, I., Wit, E.: Reproducing kernel Hilbert space based estimation of systems of ordinary differential equations. Pattern Recogn. Lett. **45**, 26–32 (2014)

Lukauskas, S., Visintainer, R., Sanguinetti, G., Schweikert, G.B.: Dgw: an exploratory data analysis tool for clustering and visualisation of epigenomic marks. BMC Bioinform. **17**(16), 53 (2016)

Macdonald, B., Higham, C., Husmeier, D.: Controversy in mechanistic modelling with gaussian processes. In: International Conference on Machine Learning, pp. 1539–1547 (2015)

Ramsay, J.O., Hooker, G., Campbell, D., Cao, J.: Parameter estimation for differential equations: a generalized smoothing approach. J. Roy. Stat. Soc.: Series B (Stat. Methodol.) **69**(5), 741–796 (2007)

Sakoe, H., Chiba, S.: Dynamic programming algorithm optimization for spoken word recognition. IEEE Trans. Acoust. Speech Signal Process. **26**(1), 43–49 (1978)

Vyshemirsky, V., Girolami, M.A.: Bayesian ranking of biochemical system models. Bioinformatics **24**(6), 833–839 (2008)

Xun, X., Cao, J., Mallick, B., Carroll, R.J., Maity, A.: Parameter estimation of partial differential equation models. J. Am. Stat. Assoc. **108**(503), 37–41 (2013). doi:10.1080/01621459.2013.794730. ISSN 0162–1459

IRIS-TCGA: An Information Retrieval and Integration System for Genomic Data of Cancer

Fabio Cumbo[1,2](✉) [iD], Emanuel Weitschek[1,3] [iD], Paola Bertolazzi[1] [iD], and Giovanni Felici[1] [iD]

[1] Institute for Systems Analysis and Computer Science,
National Research Council, Via dei Taurini 19, 00185 Rome, Italy
{fabio.cumbo,emanuel,paola.bertolazzi,giovanni.felici}@iasi.cnr.it
[2] Department of Engineering, Roma Tre University,
Via della Vasca Navale 79, 00146 Rome, Italy
[3] Department of Engineering, Uninettuno International University,
Corso Vittorio Emanuele II 39, 00186 Rome, Italy

Abstract. Data integration is one of the most challenging research topic in many knowledge domains, and biology is surely one of them. However theory and state of the art methods make this task complex for most of the small research centers. Fortunately, several organizations are focusing on collecting heterogeneous data making an easier task to design analysis tools and test biological and medical hypothesis on integrated data. One of the most evident case of such efforts is The Cancer Genome Atlas (TCGA), a data base that contains a large variety of information related to different types of cancer. This data base offers a great opportunity to those interested in performing analysis of integrated data; however, its exploitation is not so easy since non trivial efforts are required to extract and combine data before it could be analyzed in an integrated perspective. In this paper we present IRIS-TCGA, an online web service developed to perform multiple queries for data integration on TCGA. Differently from other tools that have been proposed to interact with TCGA, IRIS-TCGA allows a direct access to the data and enables to extract detailed combinations of subsets of the repository, according to filters and high-order queries. The structure of the system is simple, as it is built on two main operators, union and intersection, that are then used to construct queries of higher complexity. The first version of the system supports the extraction and integration of gene expression (RNA-sequencing, microarrays), DNA-methylation, and DNA-sequencing (mutations) data from experiments on tissues of patients, together with their related meta data, in a gene oriented organization. The extracted data matrices are particularly suited for data mining applications (e.g., classification). Finally, we show two application examples, where IRIS-TCGA is used for integrating genomic data from RNA-sequencing and DNA-methylation experiments, and where state-of-the-art bioinformatics analysis tools are applied to the integrated data in order to extract new knowledge from them. IRIS-TCGA is freely available at http://bioinf.iasi.cnr.it/iristcga/.

A. Bracciali et al. (Eds.): CIBB 2016, LNBI 10477, pp. 160–171, 2017.
DOI: 10.1007/978-3-319-67834-4_13

Keywords: Genomic data integration · Advanced queries · Knowledge extraction · Cancer · TCGA

1   Introduction

Data integration [1] is a challenging issue in analyzing data in the field of molecular biology because of two main reasons: (i) there is no one-to-one correspondence between biological entities (molecules, biological processes, etc.) and their names, and (ii) data extracted from different experiments cannot be compared due to the different physical, chemical and environmental conditions of the experiments themselves. The large amount and diversity of data produced by the new high-throughput technologies [2,3] requires an outstanding competence and domain expertise to integrate and compare all the information contained therein. Many research centers on this topic invest resources for collecting data from experiments and for the design and the development of tools for the analysis of integrated data, aiming at new insights related to diseases and drug effects in biological processes. However, most of the job to integrate data is performed by a large number of domain experts. As a consequence, it seems that data integration turns out to be an almost forbidden task for small research centers.

A positive example in this field is The Cancer Genome Atlas (TCGA) [4] (http://cancergenome.nih.gov/), a data base that contains a large variety of information related with different types of cancer. It is a large repository of genomic data extracted through controlled experiments on different tissues (samples) of patients, that overcomes the problem of experimental data comparison. This allows a large community of researchers to deal with data integration challenges. Extraction of data from TCGA and its management often requires advanced skills, as TCGA provides only a simple interface that allows the extraction of single samples and single data types at a time. To overcome these limitations, a discrete number of tools have been proposed; below we describe a subset of them:

- Anduril [5] is a software that provides the possibility to select a subset of data from TCGA, integrates, analyses, and visualizes multidimensional and heterogeneous genomic experiments and other data provided by different repositories. However, it requires strong technical skills since it is a script-based software and the user should be able to write and implement scripts to perform extraction and data analysis.
- ICGC [6], the International Cancer Genome Consortium, is a repository of research projects that include data about 50 different tumor types with clinical information about patients. Most of these data come from TCGA. For each tumor data, ICGC reports some statistics such as the percentage of male and female patients, their vital status, and tumor stage.
- GeneSpot (http://genespot.cancerregulome.org/) is a repository of knowledge extracted from TCGA genomic data. It requires advanced technical skills to retrieve data from its repository. The GeneSpot web interface includes a basic visualization tool to browse some useful statistics information for each disease.

- cBioPortal [7] is a big repository of 125 cancer studies, most of them conducted using TCGA data. It contains two data visualization tools about oncoprints generations and mutations mapping.
- TCGA-Assembler [8] is a software tool that allows the acquisition of TCGA data and the subsequent storage by transforming them in a single data table.
- TCGAbiolinks [9] is an R/Bioconductor package for the retrieval of TCGA data providing also multiple analysis methods and data visualization techniques.
- Web-TCGA [10] is an online tool that permits to integrate, analyze and visualize cancer genomic data of TCGA by defining molecular profiles in different cancer entities. Its focus is on data analysis rather than on integration.

It is worth to note that these tools focus on retrieval, assembly, analysis, and visualization of TCGA data. Conversely, IRIS-TCGA implements a set of procedures that allow to query for and extract a gene-oriented organization of experiments information, through the two simple operators of union and intersection, providing an integrated view of the data. The final results are data matrices that can be easily analyzed with data mining methods, e.g., classification.

2   Materials and Methods

In this section, we briefly present TCGA and summarize how IRIS-TCGA works as an online web tool to perform and integrate multiple queries on TCGA.

TCGA is a repository of data related to 33 cancer types. It contains results of experiments from tissues (samples) of more than 10.000 patients, and provides, for each sample, trascriptomic and genomic data, in particular RNA-transcripts expression profiles, DNA somatic mutations, and DNA methylation. Moreover, to each sample a large set of meta data is associated: clinical data of the patient (e.g., age, gender), biospecimen of the tissues (tumor stage) and information on the experiment.

IRIS-TCGA is a web tool able to search, retrieve and integrate genomic data and meta data (clinical [11], and biospecimen information) of the following experiments: RNA-sequencing [12,13], microarrays [14], DNA-methylation [15, 16], and DNA-sequencing (mutations) [17]. This first version of IRIS-TCGA deals with TCGA level 2 for DNA-sequencing and level 3 for the other types of experiments, which are high-level and pre-processed data.

IRIS-TCGA is released as an all-in-one web service that combines data retrieval and extraction with data integration, and guides the researcher through the process in a simple and intuitive way. The web interface is written in PHP (http://www.php.net) and jQuery (https://jquery.com), while the service core component is written in Java (https://www.java.com) and constantly runs in background on the server, managing all the users requests and generating the results. At present, no registration is required to use the service. The interface is simple and allows the user to extract matrices of data for a certain disease, chosen among those available in TCGA. Each matrix contains - for a set of samples possibly filtered by one or more meta data - values of one or more data types, e.g., genomic, clinical, tissue and experiment related. We chose to operate at gene level, therefore the extracted matrices are indexed on the samples and on the genes. In case a genomic experiment is performed on single sites or genomic regions, we adopt normalization and aggregation procedures to map them on the genes. Currently, DNA-methylation and DNA-sequencing (mutations) are aggregated at gene level, i.e., by considering the sum of the values of the methylated sites and the counts of the mutations in the considered gene, respectively [18]. We choose these values for the two gene-wide measures in order to see how they perform in classification and correlation tasks. See Sect. 3.2 for further details.

The extraction is performed, using three basic operations:

- **filter** data according to available meta information (e.g., age, gender, tumor grade, etc.);
- **intersect** two or more filtered data sets on samples and/or genes/transcripts;
- **unify** two or more filtered data sets on samples and/or genes/transcripts.

Filtering through one or more meta data is the simplest operation that allows to extract genomic and clinical data for a subset of samples. For an advanced usage of IRIS-TCGA, the integration operations (union, and intersection) can be applied to extract common or uncommon information from multiple data sets, e.g. RNA-transcripts expression profiles deriving from microarray and NGS technologies. If these operations are applied to the genes domain, the intersection consists of an integrated data set composed by only the genes shared by all data sets; differently, the union operation extracts all the genes that are present in the data sets, setting the values of non-common genes to zero. In case of the samples domain, the output of the intersection operation is a data set limited to samples on which the experiments corresponding to the selected genomic data sets have been performed. Indeed, the output of the union operation applied on the samples domain is limited to the samples that are present in the data sets.

Other meta data, with respect to those eventually used for filtering, can be simply added to the output matrix by selecting them from the list of available meta data.

The output produced by IRIS-TCGA can be easily given as input to data mining (i.e., clustering and classification) tools, like rule-based [11,14,19] and functions-based classifiers [20,21], see Sect. 3 for further details.

IRIS-TCGA significantly accelerates the process of searching and retrieving different types of genomic data since it operates on a local repository that is a partial mirror of TGCA data. In this first release of the web service a version of this repository has been created containing four data types - RNA-sequencing, microarrays (expression genes), DNA-methylation, and DNA-sequencing - for 15 cancer types, comprising a total of more than 5 GB of data (see Table 1 for further details). Moreover, IRIS-TCGA stores for each cancer type all meta data about the samples and patients whose tissues have been studied for that specific disease. Since IRIS-TCGA is a mirror of TCGA for a subset of genomic data, the updating of this subset is immediately produced when available.

Table 1. Volumes of genomic data available on IRIS-TCGA (in MB)

Tumor tag	Expression genes	RNA-sequencing	DNA-methylation	DNA-sequencing
BLCA	NA	49.89	174.54	3.58
BRCA	105.93	655.96	349.00	35.10
COAD	31.27	136.92	137.94	7.52
ESCA	NA	101.29	79.81	5.39
HNSC	NA	218.95	224.68	8.00
KIRC	13.02	403.86	186.38	5.08
KIRP	3.04	13.10	126.32	2.64
LAML	30.02	68.31	72.97	0.74
LIHC	NA	19.27	169.21	11.21
LUAD	5.87	121.08	196.97	7.01
LUSC	27.66	181.68	162.43	5.37
OV	107.69	165.08	4.29	0.87
READ	13.03	51.44	42.59	1.79
STAD	NA	36.92	155.23	10.05
UCEC	9.90	242.89	190.80	5.97
Total	347.43	2,466.64	2,273.16	110.32

3   Results and Discussion

In this section we describe the pipeline of IRIS-TCGA and show how to filter, extract, integrate, and download genomic and meta data.

Then, we report two application examples, where we analyze integrated data of Breast Cancer. The first one is an application of supervised learning algorithms on RNA-sequencing and DNA-methylation experiments. The second one is a linear regression analysis that aims to find correlation between three kind of experiments, i.e., RNA-sequencing, DNA-methylation, and DNA-sequencing.

3.1   Data Extraction and Integration

The working principles of IRIS-TCGA are simple (see Fig. 1). Once the user has sent a request, a process that elaborates the query and produces the expected results is launched. In order to perform a request, the user provides:

- the **disease** of interest;
- (optional) one or more **meta information** used as filters on the patient set (e.g., select all female patients only, etc.);
- at least one **data type**, selecting the platform, the TCGA data level and, the tissue (normal or tumoral);

- (optional) an **advanced integration query** applied on the set considering all their genome or vice versa, specifying union and intersection operations between the previously selected data types;
- (optional) a list of genes to which the data processing has to be restricted.

Once the user has clicked on the *Compute* button, the request is sent. For each sample the list of genes and their experimental values resulting from the execution of the advanced integration query are extracted and stored in a local flat file for each data type, containing the genes/transcripts expression values on the rows and the samples identifiers on the columns.

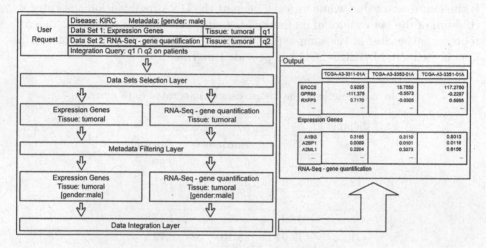

Fig. 1. Example of a IRIS-TCGA Pipeline: The figure shows an example of the IRIS-TCGA work flow. Once the user has submitted her request, considering the KIRC disease, all experiments for each data type (e.g., Expression Genes, and RNA-sequencing - gene quantification) are retrieved. Then, information for all male patients samples only are filtered out. The integration query is then applied, generating a table for each data set: in this specific example, all tables contain different genomic information for the same sample.

3.2 Classification Analysis

In order to show the usage of IRIS-TCGA, we perform an integration query on genomic data of Breast Cancer. Specifically, we use IRIS-TCGA to extract RNA-sequencing and DNA-methylation experimental data for performing an integration query (intersection) on the samples. It is worth noting that, the engine of IRIS-TCGA executes an indexing on the genes. Therefore, we obtain data matrices, whose columns represents measures on a given gene. In case of RNA-sequencing, we already have the gene expression value represented with the wide-spread RPKM (Reads Per Kilobase per Million mapped reads) value [12]. Conversely, when dealing with DNA-methylation data IRIS-TCGA has to calculate a gene wide measure, because the methylation values in TCGA refer to single sites of the genome. TCGA uses the beta value [22] to measure the quantity of methylated molecules in a site. As gene wide measure we adopt the sum of the beta values of the methylated sites that are present in the gene. The same approach is used in [18], where the authors show the capability of this gene wide measure to distinguish tumoral from non-tumoral samples by applying supervised classification methods. Finally, we obtain a data matrix, whose columns represent the samples and whose rows are associated to all genes of the RNA-sequencing experiment and all genes of the DNA-methylation experiment (genes present in both experiments are duplicated). These are the features of the classification problem. For each gene of the RNA seq experiments the entry is the expression value, while for each gene of the DNA-methylation the entry is the sum of the beta values of its methylated sites. The second row of the matrix represents the class of the sample (i.e., tumoral, normal). An example of a data matrix extracted by IRIS-TCGA is provided in Table 2

Table 2. Example of the breast cancer integrated data matrix.

Sample Id	S_1	S_2	$\cdots$	S_n
Class	Tumoral	Normal	$\cdots$	Tumoral
$G1_{rna}$	1.24	6.21	$\cdots$	2.57
$G2_{rna}$	5.30	3.74	$\cdots$	4.01
$\cdots$	$\cdots$	$\cdots$	$\cdots$	$\cdots$
Gk_{rna}	3.20	2.15	$\cdots$	1.37
$G1_{dmet}$	12.04	13.10	$\cdots$	8.72
$G2_{dmet}$	5.23	1.97	$\cdots$	3.16
$\cdots$	$\cdots$	$\cdots$	$\cdots$	$\cdots$
Gm_{dmet}	1.53	6.53	$\cdots$	2.15

The extracted breast cancer data is composed of 489 tumoral samples, 74 normal samples, and 41.124 genes.

The obtained data matrix can be easily analyzed with supervised learning algorithms [23], i.e. classification. We focus on classification algorithms that compute a meaningful model, in order to identify the genes that are related with the disease under study. Specifically, we decide to adopt rule-based classifiers [24], whose classification model is composed of "if-then rules", e.g., "if the expression of ABA < 1.9 then the sample can be classified as tumoral". Recently, a new rule-based classification algorithm - CAMUR - [25], which is able to extract multiple and equivalent classification models (and therefore more knowledge), is released. We adopt CAMUR to perform the classification analysis on the integrated data set of Breast Cancer extracted by IRIS-TCGA. We run the program by considering 50 iterations that lead to more than 100 classification models composed of 200 genes. Specifically, we obtained 102 equivalent rule-based classification models with an average f-measure (accuracy) of 0.97 (0.99) when applying a holdout validation (random percentage split of 80% for training and 20% for testing). The rules are compact (i.e., the number of literals are less than six) and include combinations of genes derived from the DNA-methylation and RNA-sequencing experiments. Examples of classification rules are provided in Table 3. It is worth noting that, both combinations of genes derived from the RNA-sequencing and DNA-methylation experiments are present in the rules.

Table 3. A sample of classification rules extracted by CAMUR.

$(TMEM220_{rseq} \geq 2.5)$ AND $(LIMS2_{rseq} \geq 10.7)$ OR $(BGN_{rseq} \leq 30.4)$ AND $(HTR4_{dmeth} \geq 7.4)$
$(SDPR_{rseq} \geq 12.330)$ AND $(ANXA1_{rseq} \geq 161.3)$ OR $(B3GALT1_{rseq} \geq 0.6)$ AND $(HTR4_{dmeth} \geq 8.2)$
$(TMEM220_{rseq} \geq 2.5)$ AND $(LIMS2_{rseq} \geq 10.7)$ OR $(OR6C65_{rseq} \geq 0)$ AND $(HTR4_{dmeth} \geq 7.4)$
$(TMEM220_{rseq} \geq 2.6)$ AND $(GNG11_{rseq} \geq 32.9)$ OR $(MYH11_{rnaseq} \geq 75.5)$ AND $(CRHBP_{dmeth} \geq 8.1)$
$(CD300LG_{rseq} \geq 10.3)$ AND $(FAM54B_{rseq} \geq 20.4)$ OR $(GRAMD3_{rseq} \geq 31)$ AND $(HTR7_{dmeth} \geq 3)$
$(FMO2_{rseq} \geq 15.2)$ AND $(TPO_{rseq} \geq 0.3)$ AND $(SHMT1_{rseq} \geq 11)$ OR $(GCDH_{dmeth} \leq 1.6)$
$(PPAP2B_{rseq} \geq 38.1)$ AND $(CDKN1C_{rseq} \geq 8.7)$ OR $(KCNE1_{rseq} \geq 1.9)$ AND $(TNFRSF4_{dmeth} \geq 7.6)$
$(CHL1_{rseq} \geq 3.8)$ AND $(SLC35A2_{rseq} \leq 13)$ OR $(TCEAL7_{rseq} \geq 5.5)$ AND $(FAM19A2_{dmeth} \geq 9.6)$
$(NKAPL_{rseq} \geq 0.5)$ AND $(HIST1H3J_{rseq} \leq 0.1)$ OR $(MYH11_{rseq} \geq 58.5)$ AND $(LOC643719_{dmeth} \leq 5.4)$
$(FXYD1_{rseq} \geq 19.2)$ OR $(OXTR_{rseq} \geq 30.2)$ AND $(ORAI1_{dmeth} \geq 4.4)$

The genes that are present in the models can be further investigated by domain experts by performing wide-spread bioinformatics analyses, e.g., functional enrichment analysis [26] and pathway analysis [27].

3.3   Linear Regression Analysis

Another application example of IRIS-TCGA is described in this subsection. The problem of defining if there is correlation between DNA-methylation and gene expression is widely investigated in literature. Specifically, several studies try to find out if there is correlation between the methylated sites of a gene and its expression value [28–31]. Our experiment instead wishes to prove if there is correlation among the defined gene wide measure of DNA-methylation (sum of the beta value of the methylated sites for each gene) and the RPKM values, focusing on those genes that present mutations. In particular, we extract genomic data of Breast Cancer by considering RNA-sequencing, DNA-methylation, and

DNA-sequencing experiments indexed on the genes. We consider only the genes for which at least a mutation occurs. An integration (intersection) of the samples is performed obtaining a data matrix whose columns represent the samples and whose rows the genes replicated for RNA-sequencing and DNA-methylation experiments. The entries contain the experimental values, i.e., RPKM for RNA-sequencing, and the sum of the beta values for DNA-methylation. It is worth to note that the final output comprises only the samples and the genes that are in common between the different genomic experiments. Therefore the data set is composed of 468 samples and 1592 genes.

Aim of our analysis is to investigate if the RNA-sequencing RPKM value of each gene is related to the DNA-methylation experiments. We apply a linear regression model [32] in order to show the correlation among the different genomic experiments. The regression is performed for each gene, where we consider the R2 value, i.e., a measure how the RPKM gene expression is correlated with the DNA-methylation, and the mutations. For further details the reader may refer to Fig. 2, where we draw the R2 value for each considered gene. Our results show that a correlation exists only for a few set of genes, which can be further analyzed by domain experts. This fact is also confirmed by previous studies on different cancer types (e.g., [28–31]).

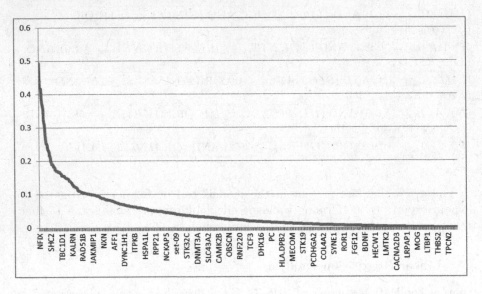

Fig. 2. Linear regression values among RNA-sequencing and DNA-methylation calculated for each gene that presents mutations.

4   Conclusions

In this work we described IRIS-TCGA an online tool to query, retrieve, filter and produce an integrated view of genomic data of cancer from TCGA. IRIS-TCGA integration capabilities are based on two simple operators (i.e., union and intersection), which work both on the samples and on the genes present in the genomic experiments. Currently, it supports gene expression (RNA-sequencing and microarrays), DNA-methylation, and DNA-sequencing (mutation) data and their related meta data. Two application examples have been described to show how IRIS-TCGA is able to integrate and extract data from TCGA and to create in a very simple way the input to data mining tools and statistical routines, like linear regression.

Future work will include the following aspects. First, DNA-methylation and DNA-sequencing characteristics will be investigated to derive other proper query and integration routines to obtain the correct formats for different analysis tasks. Then, thanks to its modular structure, it is easy to extend the functions of IRIS-TCGA in order to integrate other types of data such as Copy Number Variations [33] and microRNA sequencing (miRNA-seq) [34]. Additionally, we plan to define new integration functions that permit the extraction of non aggregated data. Moreover, the integration of different genomic open access data bases, and the use of IRIS-TCGA to perform comparative analyses on integrated data of cancer is planned.

Finally, it is worth to note that IRIS-TCGA is already integrated in the Galaxy framework [35] as a data source tool.

Acknowledgments. The results reported here are based upon the data generated by the TCGA Research Network: http://cancergenome.nih.gov/.

Funding. The work was financially supported by the SysBioNet, Italian Roadmap Research Infrastructure, and the Epigenomics Flagship Project EPIGEN [PB.P01].

References

1. Gomez-Cabrero, D., Abugessaisa, I., Maier, D., Teschendorff, A., Merkenschlager, M., Gisel, A., Ballestar, E., Bongcam-Rudloff, E., Conesa, A., Tegnér, J.: Data integration in the era of omics: current and future challenges. BMC Syst. Biol. 8(Suppl 2), I1 (2014)
2. Hayden, E.C.: Technology: the $1,000 genome. Nature 507(7492), 294–5 (2014)
3. Weitschek, E., Santoni, D., Fiscon, G., De Cola, M.C., Bertolazzi, P., Felici, G.: Next generation sequencing reads comparison with an alignment-free distance. BMC Res. Notes 7(1), 869 (2014)
4. Weinstein, J.N., Collisson, E.A., Mills, G.B., Shaw, K.R.M., Ozenberger, B.A., Ellrott, K., Shmulevich, I., Sander, C., Stuart, J.M., Network, T.: The cancer genome atlas pan-cancer analysis project. Nature Genet. 45(10), 1113–1120 (2013)
5. Ovaska, K., Laakso, M., Haapa-Paananen, S., Louhimo, R., Chen, P., Aittomaki, V., Valo, E., Nunez-Fontarnau, J., Rantanen, V., Karinen, S., et al.: Large-scale data integration framework provides a comprehensive view on glioblastoma multiforme. Genome Med. 2(9), 65 (2010)
6. Joly, Y., Dove, E.S., Knoppers, B.M., Bobrow, M., Chalmers, D.: Data sharing in the post-genomic world: the experience of the international cancer genome consortium (ICGC) data access compliance office (daco). PLoS Comput. Biol. 8(7), e1002549 (2012)

7. Cerami, E., Gao, J., Dogrusoz, U., Gross, B.E., Sumer, S.O., Aksoy, B.A., Jacobsen, A., Byrne, C.J., Heuer, M.L., Larsson, E., et al.: The CBIO cancer genomics portal: an open platform for exploring multidimensional cancer genomics data. Cancer Disc. **2**(5), 401–404 (2012)

8. Zhu, Y., Qiu, P., Ji, Y.: TCGA-assembler: open-source software for retrieving and processing TCGA data. Nat. Methods **11**(6), 599–600 (2014)

9. Colaprico, A., Silva, T.C., Olsen, C., Garofano, L., Cava, C., Garolini, D., Sabedot, T.S., Malta, T.M., Pagnotta, S.M., Castiglioni, I., et al.: TCGAbiolinks: an R/Bioconductor package for integrative analysis of TCGA data. Nucl. Acids Res. **44**, e71 (2015)

10. Deng, M., Brägelmann, J., Schultze, J.L., Perner, S.: Web-TCGA: an online platform for integrated analysis of molecular cancer data sets. BMC Bioinform. **17**(1), 1 (2016)

11. Weitschek, E., Felici, G., Bertolazzi, P.: Clinical data mining: problems, pitfalls and solutions. In: 24th International Workshop on Database and Expert Systems Application, pp. 90–94, 10662 Los Vaqueros Circle, Los Alamitos, CA 90720, USA. IEEE Computer Society (2013)

12. Mortazavi, A., Williams, B.A., McCue, K., Schaeffer, L., Wold, B.: Mapping and quantifying mammalian transcriptomes by RNA-Seq. Nat. Methods **5**(7), 621–628 (2008)

13. Li, B., Dewey, C.N.: RSEM: accurate transcript quantification from RNA-Seq data with or without a reference genome. BMC Bioinform. **12**(1), 323 (2011)

14. Weitschek, E., Felici, G., Bertolazzi, P.: Mala: a microarray clustering and classification software. In: 23rd International Workshop on Database and Expert Systems Application, pp. 201–205, 10662 Los Vaqueros Circle, Los Alamitos, CA 90720, USA. IEEE Computer Society (2012)

15. Bird, A.P.: CpG-rich islands and the function of DNA methylation. Nature **321**(6067), 209–213 (1985)

16. Bird, A.: DNA methylation patterns and epigenetic memory. Genes Dev. **16**(1), 6–21 (2002)

17. McKenna, A., Hanna, M., Banks, E., Sivachenko, A., Cibulskis, K., Kernytsky, A., Garimella, K., Altshuler, D., Gabriel, S., Daly, M., et al.: The genome analysis toolkit: a mapreduce framework for analyzing next-generation DNA sequencing data. Genome Res. **20**(9), 1297–1303 (2010)

18. Weitschek, E., Cumbo, F., Cappelli, E., Felici, G.: Genomic data integration: a case study on next generation sequencing of cancer. In: 27th International Workshop on Database and Expert Systems Application, pp. 49–53, 10662 Los Vaqueros Circle, Los Alamitos, CA 90720, USA. IEEE Computer Society (2016)

19. Cohen, W.W.: Fast effective rule induction. In: Proceedings of the Twelfth International Conference on Machine Learning, pp. 115–123. Morgan Kaufmann (1995)

20. Bishop, C.M.: Neural Networks for Pattern Recognition. Oxford University Press, New York (1995)

21. Cristianini, N., Shawe-Taylor, J.: An Introduction to Support Vector Machines and Other Kernel-Based Learning Methods. Cambridge University Press, Cambridge (2000)

22. Bibikova, M., Barnes, B., Tsan, C., Ho, V., Klotzle, B., Le, J.M., Delano, D., Zhang, L., Schroth, G.P., Gunderson, K.L., et al.: High density dna methylation array with single cpg site resolution. Genomics **98**(4), 288–295 (2011)

23. Weitschek, E., Fiscon, G., Felici, G.: Supervised DNA Barcodes species classification: analysis, comparisons and results. BioData Mining **7**(1), 1 (2014)

24. Tan, P., Steinbach, M., Kumar, V.: Introduction to Data Mining. Addison Wesley, Boston (2005). 75 Arlington Street, Suite 300
25. Cestarelli, V., Fiscon, G., Felici, G., Bertolazzi, P., Weitschek, E.: CAMUR: Knowledge extraction from RNA-seq cancer data through equivalent classification rules. Bioinformatics 32(5), 697–704 (2016)
26. Dandrea, D., Grassi, L., Mazzapioda, M., Tramontano, A.: Fidea: a server for the functional interpretation of differential expression analysis. Nucl. Acids Res. 41(W1), W84–W88 (2013)
27. Khatri, P., Sirota, M., Butte, A.J.: Ten years of pathway analysis: current approaches and outstanding challenges. PLoS Comput. Biol. 8(2), e1002375 (2012)
28. Kulis, M., Heath, S., Bibikova, M., Queirós, A.C., Navarro, A., Clot, G., Martínez-Trillos, A., Castellano, G., Brun-Heath, I., Pinyol, M., et al.: Epigenomic analysis detects widespread gene-body DNA hypomethylation in chronic lymphocytic leukemia. Nature Genet. 44(11), 1236–1242 (2012)
29. Chen, C., Zhang, C., Cheng, L., Reilly, J.L., Bishop, J.R., Sweeney, J.A., Chen, H.Y., Gershon, E.S., Liu, C.: Correlation between DNA methylation and gene expression in the brains of patients with bipolar disorder and schizophrenia. Bipolar Disorders 16(8), 790–799 (2014)
30. Akalin, A., Garrett-Bakelman, F.E., Kormaksson, M., Busuttil, J., Zhang, L., Khrebtukova, I., Milne, T.A., Huang, Y., Biswas, D., Hess, J.L., et al.: Base-pair resolution DNA methylation sequencing reveals profoundly divergent epigenetic landscapes in acute myeloid leukemia. PLoS Genet. 8(6), e1002781 (2012)
31. Maunakea, A.K., Nagarajan, R.P., Bilenky, M., Ballinger, T.J., DSouza, C., Fouse, S.D., Johnson, B.E., Hong, C., Nielsen, C., Zhao, Y., et al.: Conserved role of intragenic DNA methylation in regulating alternative promoters. Nature 466(7303), 253–257 (2010)
32. Seber, G.A., Lee, A.J.: Linear Regression Analysis, vol. 936. Wiley, Hoboken (2012). 07030–5774
33. Conrad, D.F., Pinto, D., Redon, R., Feuk, L., Gokcumen, O., Zhang, Y., Aerts, J., Andrews, T.D., Barnes, C., Campbell, P., et al.: Origins and functional impact of copy number variation in the human genome. Nature 464(7289), 704–712 (2010)
34. Zeng, Y., Cullen, B.R.: Sequence requirements for micro RNA processing and function in human cells. RNA 9(1), 112–123 (2003)
35. Blankenberg, D., Kuster, G.V., Coraor, N., Ananda, G., Lazarus, R., Mangan, M., Nekrutenko, A., Taylor, J.: Galaxy: a web-based genome analysis tool for experimentalists. Current Protocols Mol. Biol. 19, 1–21 (2010)

Effect of UV Radiation on DPPG and DMPC Liposomes in Presence of Catechin Molecules

Filipa Pires, Gonçalo Magalhães-Mota, Paulo António Ribeiro,
and Maria Raposo[✉]

Department of Physics, Faculdade de Ciências e Tecnologias, FCT,
Centre of Physics and Technological Research, CEFITEC,
Universidade Nova de Lisboa, Caparica, Portugal
{af.pires,g.barreto}@campus.fct.unl.pt, {pfr,mfr}@fct.unl.pt

Abstract. Catechin molecules are known to reduce the oxidative stress-induced by radiation acting as scavenger of the reactive oxygen species, preventing in this way the damage in biomolecules. In this work, the effect of radiation on liposomes of 1,2-dipalmitoyl-sn-glycero-3-[phospho-rac-(1-glycerol)(sodium salt) (DPPG) and of 1,2-dimyristoyl-sn-glycero-3-phosphocholine (DMPC) is analyzed in the absence and presence of epigallocatechin-3-gallate (EGCG) molecules, having in view the evaluation of the photosensitizing properties and the efficacy of these molecules to modulate cell membrane damage mechanisms. The obtained results demonstrate that the damage by UV radiation on DPPG and DMPC liposomes is strongly dependent of the presence of EGCG molecules. While DPPG liposomes are protected from radiation in presence of EGCG, the EGCG molecules are damaged by the radiation supporting the idea that EGCG are strongly adsorbed on the inner and outer liposome surfaces due hydrogen bonding. This suggests that EGCG molecules in the inner surface can be protected from radiation. In the case of DMPC liposomes, the EGCG molecules are affected by radiation as well as the DMPC molecules. This is explained if the EGCG chroman group is positioned between DMPC lipids while the gallic acid groups float over the liposomes.

Keywords: Cell membrane · Natural antioxidant · Physical interactions · Cellular detoxification · Delivery system

1 Introduction

The addition of natural antioxidants to diet is belived to attenuated and even promote the repair of some DNA lesions, through the regulation of DNA methylation involved-genes as well as the modulation of intracellular redox environment [1]. Currently, epidemiological studies had recommend the intake of catechins, namely epigallocatechin-3-gallate (EGCG), since these dietary molecules are strong enough to modulate the intracellular signaling pathways involved in the regulation of apoptosis, angiogenesis and metastasis, preventing in this way

© Springer International Publishing AG 2017
A. Bracciali et al. (Eds.): CIBB 2016, LNBI 10477, pp. 172–183, 2017.
DOI: 10.1007/978-3-319-67834-4_14

cancer progression [2,3]. Catechins, a flavonoid subclasse, are found in tea, one of the most consumed beverage in the world after water [4]. Catechins uptake by cellular machinery occurs by passive transport, being their biological activity governed by their affinity for lipid bilayer [5]. However, the effectiveness of catechin in pharmaceutical applications is frequently constrained by their vulnerability to environmental factors as temperature, pH and humidity, leading to their epimerization, oxidation or degradation. To avoid this, catechins should be encapsulated in some drug delivery systems, as for example liposomes.

Liposomes are in fact delivery systems that can protect catechins from the extreme environmental conditions mentioned above, allowing a more efficient delivery and release of these antioxidants to cells. The study of action mode, at molecular level, of these dietary-derived antioxidant molecules, such as catechins, can lead to new approaches towards DNA repair and cancer prevention.

Nowadays, computational tools have been addressed to identify new molecular targets of EGCG and to reveal the interactions mechanisms (van der Waals, hydrogen bonding) ruling chemical reactions [6,7]. These tools provide insights about molecular targets of EGCG, but do not exclude experimental studies of the interactions that molecules such as catechins can make with biological molecules or membrane structures that can be easily mimicked by liposomes.

Flavonoids also have attracted considerable interest due to their ability to attenuate the oxidative stress induced by radiation. In fact, ultraviolet radiation is a powerful environmental agent which strongly interacts with biological molecules, namely the deoxyribonucleic acid (DNA), inducing single and double strand breaks (SSBs and DSBs) [8]. For example, data obtained from DNA thin films UV exposed for different irradiation times revealed the decrease of the number of thymines involved in Hoogsteen base pairing with adenine and of the number of phosphate groups, as a result of the opening of sugar chains [9]. In addition, AC electrical conductivity measurements carried out on DNA thin films, irradiated with UV light revealed that electrical conduction is arising from DNA chain electron hopping, between base-pairs and phosphate groups [10], being the hopping distance calculated from correlated barrier hopping model, equal to the distance between DNA base-pairs. These results also allowed to conclude that the loss in conductivity with irradiation time was arising from the decrease of the phosphates groups in DNA molecules. Further analysis of the damage at the molecular level caused by UV light in the range from 3.5 to 8 eV on DNA films allowed to conclude that the damage on bases peripheral nitrogen atoms follows that of the phosphates. This suggests that very low energy photoelectrons are ejected from the DNA bases, as a result of UV light induced breaking of the phosphate ester groups, given rise to a transient anion and resonance formation with removal of the nitrogen DNA peripheral groups [11]. The above results described lead us to admit that it would be difficult for cells to repair from the breaks generated by UV, fact that will contribute to mutations and, consequently, to diseases. Different research areas have been debating and producing new tools, having in view the understanding of the molecular mechanisms underlying the harmful agents activity in cellular machinery.

Cell membrane is a complex barrier formed by charged-lipids in the inner leaflet and by glycolipids in the outer leaflet. Their complexity and dynamic nature makes extremely difficult analyze the biophysical interactions between drugs and cell membrane namely with respect to selective transport or uptake of the drug by intracellular targets. Physicochemical properties of drugs as polarity, charge, molecular weight, solubility and pH greatly influence the efficiency of their transport across the membrane. Catechins uptake by the cellular machinery occurs by passive transport, being their biological activity governed by their lipid bilayer affinity. A higher hydrophobicity as it can be inferred from see the EGCG chemical structure in the inset of Fig. 1, ensure more interaction of catechins with lipid bilayers, meaning that the cellular uptake of gallate catechins (EGCG) will be higher than in non-gallate catechins. As the catechins intake can attenuate or even repair the biomolecules damage via antioxidant mechanisms or by modulating the intracellular redox environment, the main objective of this work was to analyze the effect of UV radiation on catechin molecules as epigallocatechin-3-gallate (EGCG) incorporated on liposomes of 1,2-dipalmitoyl-sn-glycero-3-[phospho-rac-(1-glycerol)(sodium salt) (DPPG) and of 1,2-dimyristoyl-sn-glycero-3-phosphocholine DMPC, in order to understand how catechins interact with cellular membranes and their ability to attenuate the damage induced by radiation. In this work, in addition to study the protective role of EGCG, we also want to study the possibility of use DPPG and DMPC liposomes as stable vehicles to encapsulate EGCG, aspiring the use of these liposomes in cancer therapy. Saturated lipids as DPPG were considered in this study, since DPPG is one of the main components of the mammalian pulmonary surfactant and the development of new liposomal formulations carrying EGCG, offers a possibility to increase the bioavailability and favor intracellular delivery of EGCG in lung tissue to suppress the growth of lung cancer cells. In relation to DMPC, we decided to study this saturated lipid, since some tumors contain high levels of saturated phosphatidylcholines and, few studies focused on the link between the misbalance of saturated/unsaturated fatty acids metabolism and cancer metabolism.

2   Materials and Methods

Epigallocatechin-3-gallate (EGCG) was acquired Sigma-Aldrich while the synthetic phospholipids 1,2-dipalmitoyl-sn-glycero-3-[phospho-rac-(1-glycerol)(sodium salt) (DPPG)or 1,2-dimyristoyl-sn-glycero-3-phosphocholine (DMPC) were supplied by Avanti Polar Lipids (Alabaster, AL). Liposomes containing 5 mM of phospholipids were prepared by dry film method [12–14], where the phospholipids were dissolved in a minimum amount of (8:2) chloroform:methanol. The organic solvent was evaporated using a soft nitrogen flow to form a lipid thin film in the walls of a falcon tube. The film was then hydrated with ultrapure water, obtained from Milli-Q water system, and with an aqueous solution containing 400 µM of EGCG. An ultrasonic processor UP50H (Dr. Hielscher GmbH, Germany) was used to sonicate the obtained liposome suspension in an ice bath, for 30 s at 50 W. This procedure was repeated 15 times with 1 min rest time in each cycle. The DPPG suspension with and without EGCG were exposed to a 254 nm UVC germicide lamp with a radiance of $1.9 \, W/m^2$. All the UV irradiated emulsions present a lipidic concentration of 0.05 mM and the concentration of EGCG in the DPPG+EGCG and DMPC+EGCG emulsions was of $4 \, \mu M$. The concentration of EGCG solution was of $4 \, \mu M$. Solutions/emulsions were characterized by a spectrophotometer Shimadzu, model UV-2101.

3 Results

3.1 Effect of Ultraviolet Radiation on Catechin Molecules in Aqueous Solutions

To analyze the behavior of catechin molecules when subjected to UV radiation, aqueous solutions of EGCG were irradiated and the UV-visible spectra of these solutions were measured for different irradiation times. The obtained UV-Vis spectra are shown in Fig. 1. Spectra present absorbance bands at 206 nm and 276 nm, corresponding to electronic $\pi - \pi^*$ and $n - \pi^*$ transitions in which chroman group and phenol (gallic acid group) oxygen electrons are involved [15,16]. The intensity of these bands is seen to decrease quasi exponentially with the irradiation time, as shown in Fig. 1(b). The obtained characteristics time values were of 48 ± 8 and 90 ± 30 min, respectively, indicating that damage induced in EGCG molecules is likely due to independent processes.

3.2 Effect of Ultraviolet Radiation on DPPG Liposomes

The influence of UV exposure on DPPG emulsion in the absence and in the presence of EGCG was analyzed by measuring UV-Vis absorbance spectra for different exposure times. These results are shown in Fig. 2(a) and (b), respectively. Previous characterization of vacuum ultraviolet (VUV) spectra of DPPG cast films [13] revealed a presence of a band at 194.4 ± 0.7 nm (6.38 ± 0.04) eV which has been assigned either to the $n_0 \rightarrow \pi^*$ [17,18] transition from the lone-pair on the carbonyl oxygen to the antibonding CO valence orbital and to the valence shell electronic excitations of hydroxyl groups [19–21]. In the spectra of Fig. 2(a), one can observe a shoulder of a peak at a wavelength below 194 nm. A plausible explanation for that is a hypochromic shift of the band since phospholipids are suspended in water and not immobilized in a solid substrate. Another explanation is the resolution of our apparatus compromised the observation of the bands near of 200 nm. The UV exposure leads to the appearance of a new band at 215 nm, whose absorbance intensity increases exponentially with irradiation time, as plotted in Fig. 3. This behavior can be due to the UV light inducing the formation of reactive oxygen species and damage the phosphate heads of phospholipid, resulting in a great number of oxygens which have a high affinity to carbon in the tails of phospholipids forming new ether bonds, which strongly absorb at 215 nm. Furthermore, the UV light can also promote the breakage of

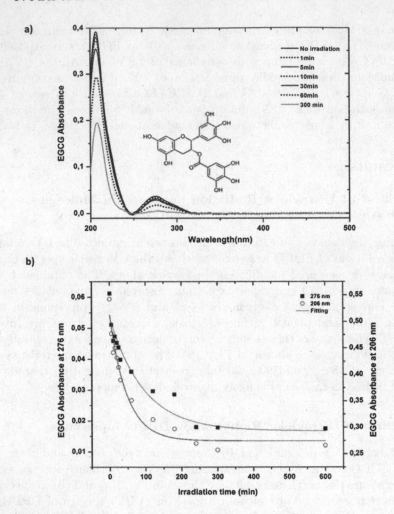

Fig. 1. (a) Absorbance spectra of EGCG aqueous solutions subjected to UV radiation during different periods of time. In the inset is shown the molecular structure of EGCG. (b) Evolution of absorbance at 206 and at 276 nm of EGCG aqueous solutions as a function of the irradiation time. Solid lines are guidelines.

carbons tails, leading to the formation of acyclic conjugated dienes, as in the case of unsaturated lipids, which also strongly absorb in the region of 215–233 nm. The presence of this new band is consistent with lipid oxidation phenomenon taking place. In presence of catechin molecules, the absorbance at 215 nm of the DPPG+EGCG emulsion decrease with the irradiation time following the exponential plot curve as also shown in plot of Fig. 3. In this figure it was also included the plot of the absorbance at 215 nm for the irradiated EGCG aqueous solutions, see Fig. 1, to better see the effect of radiation. It should be also referred that the absorbance at 215 nm in the DPPG+EGCG emulsion spectra is due to EGCG

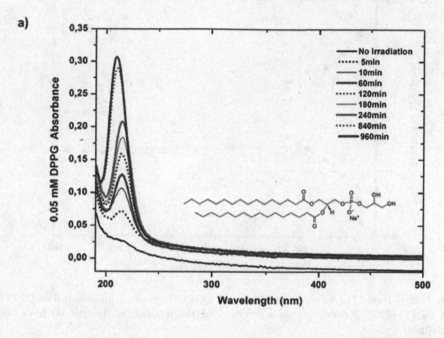

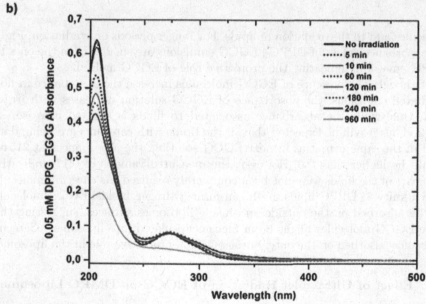

Fig. 2. Absorbance spectra of (a) DPPG and (b) DPPG+EGCG emulsions subjected to UV radiation during different periods of time. In the inset of (a) is shown the molecular structure of DPPG phospholipid.

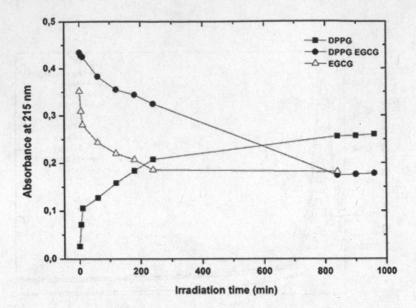

Fig. 3. Evolution of absorbance at 215 nm of EGCG aqueous solutions and of DPPG and DPPG+EGCG emulsions as a function of the irradiation time. Solid lines are guidelines.

molecules and to the oxidation of lipids. For longer periods of irradiation, Fig. 3, the absorbance values of DPPG+EGCG emulsions are smaller than the ones for DPPG emulsion, indicating the protective role of EGCG molecules.

Although, the presence of EGCG molecules increase the absorbance in non-irradiated emulsions, the absorbance of EGCG solution decreases with irradiation time while the absorbance associated to lipids is seen to increase, see Fig. 2. Thus it will be expected that, if the lipids with catechins are being damaged in the same extent as in neat EGCG solution, the absorbances at 215 nm should be higher than 0.3. However, the measured value is of 0.17, indicating that part of the lipids were not been completely oxidized. As catechin molecules have affinity to DPPG lipids, both can form hydrogen bonds, EGCG molecules can be adsorbed on the outside and inside liposomes surfaces, suggesting that some EGCG molecules might be in fact encapsulated into liposomes. Catechin molecules adsorbed on the outer surface have a protective role in the liposomes.

3.3 Effect of Ultraviolet Radiation of EGCG on DMPC Liposomes

Taking into account the molecular structure of DMPC molecules, the number of hydrogen bonds that DMPC molecules can do with EGCG molecules, should be reduced when compared with the case of DPPG molecules. In order to better understand what is happening when EGCG molecules are close to DMPC liposomes in presence of UV radiation, DMPC and DMPC+EGCG emulsions

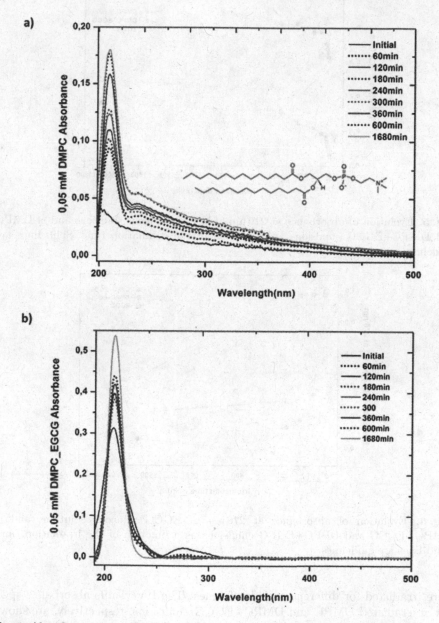

Fig. 4. Absorbance spectra of (a) DMPC and (b) DMPC+EGCG emulsions subjected to UV radiation during different periods of time. In the inset of (a) is shown the molecular structure of DMPC phospholipid.

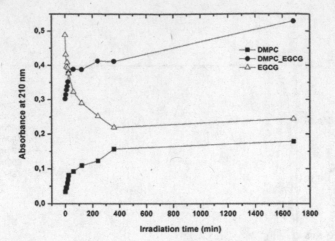

Fig. 5. Evolution of absorbance at 210 nm of EGCG aqueous solutions and of DMPC and DMPC+EGCG emulsions as a function of the irradiation time. Solid lines are guidelines.

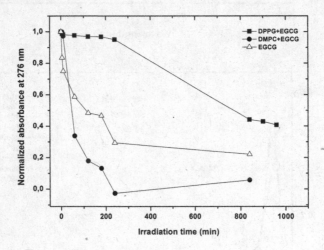

Fig. 6. Evolution of absorbance at 276 nm of EGCG aqueous solutions and of DMPC+EGCG and DPPG+EGCG emulsions as a function of the irradiation time. Solid lines are guidelines.

were irradiated for different periods of times. The UV-visible absorbance spectra of irradiated DMPC and DMPC+EGCG emulsions, respectively, are shown in Figs. 4(a) and (b). As had been observed for DPPG, also the DMPC molecules are being damaged when irradiated. In fact, the UV-visible spectra show an increase of the absorbance intensity of the band at 210 nm and of the broad band at 238 nm. Nonetheless, in presence of EGCG molecules the 238 nm broad band does not appear at all, while the band at 210 nm is seen to increase in intensity as can be seen in Fig. 5, where absorbance at 210 nm is plotted as a function of irradiation time. This indicates that DMPC molecules are being oxidized. Also, the EGCG molecules are being damaged since the intensity of EGCG peak at 276 nm decreases with the irradiation time. To better understand the behavior of EGCG molecules close to lipids, the normalized absorbance at 276 nm associated with the EGCG band was plotted as a function of irradiation time in Fig. 6. Results show that in the presence of DPPG, the 276 nm absorbance band disappears at a lower rate than in the case of EGCG+DMPC, suggesting that EGCG molecules are encapsulated by DPPG liposomes. On the other hand, since the 276 nm band flatens rapidly and the DMPC 238 nm broad band is not coming with the irradiation time, some chemical groups of EGCG are likely to bound to DMPC liposomes. This result can be explained if some chemical groups of EGCG are being bound to DMPC liposomes contributing for the liposome membrane rigidification. This phenomenon was already reported [22]. More recently, Phan et al. demonstrated that flavonoids as EGCG caused lipid membrane aggregation and rigidification. Due to the presence of gallate, galloyl and hydroxyl groups, flavonoids were able to form hydrogen bonds with head group of membrane phospholipids. It was demonstrated that they mainly affect the hydrophilic region of lipid bilayers, mediated phospholipid aggregation, thus causing a decrease in membrane area and rendering a membrane more rigid [23]. These authors proposed that the chroman group is able to stay between lipid headgroups while the gallic acid groups are left out of liposome. In fact, such arrangement can explain the present experimental results. If the chroman groups are displaced between the lipid headgroups they can be protected from radiation, while the gallic acid are exposed to radiation and can be damaged.

4   Conclusion

The influence of UV radiation on DPPG and DMPC liposomes is strongly dependent on the presence of EGCG molecules. It was demonstrated that DPPG liposomes are being protected from radiation in presence of EGCG, while EGCG molecules are being damaged, supporting the idea that EGCG are strongly adsorbed at the inner and outer liposome surfaces through hydrogen bonding. The EGCG molecules bound at the inner liposomes surface can be protected from radiation. In the case of DMPC liposomes, although hydrogen bonds or hydrophobic interactions are taking place between DMPC and EGCG molecules, both EGCG as well DMPC molecules are affected by radiation. This is possible if the EGCG chroman group is positioned between DMPC lipids, while the gallic acid groups float over the liposomes. Therefore, liposomes can be suitable delivery systems to overcome the instability of EGCG, if one chooses the adequate lipid.

Acknowledgments. This work was supported by the Portuguese research Grant UID/FIS/00068/2013 by the project PTDC/FIS-NAN/0909/2014 through FCT-MCTES, Portugal. Filipa Pires acknowledges the fellowship PD/BD/106036/2015 from RABBIT Doctoral Programme (Portugal).

References

1. Lee, W.J., Shim, J.-Y., Zhu, B.T.: Mechanisms for the inhibition of DNA methyl-transferases by tea catechins and bioflavonoids. Mol. Pharmacol. **68**, 1018–1030 (2005)

2. Shankar, S., Ganapathy, S., Hingorani, S.R., Srivastava, R.K.: EGCG inhibits growth, invasion, angiogenesis and metastasis of pancreatic cancer. Front. Biosci. **13**, 440–452 (2007). A journal and virtual library

3. Isemura, M., et al.: Tea catechins and related polyphenols as anti-cancer agents. Biofactors **13**, 81–85 (2000)

4. Ross, J.A., Kasum, C.M.: Dietary flavonoids: bioavailability, metabolic effects, and safety. Ann. Rev. Nutr. **22**, 19–34 (2002)

5. Ottova-Leitmannova, A.: Advances in Planar Lipid Bilayers and Liposomes, vol. 4. Elsevier/Academic Press (2006)

6. Santos, H.A., Vila-Vicosa, D., Teixeira, V.H., Baptista, A.M., Machuqueiro, M.: Constant-ph MD simulations of DMPA/DMPC lipid bilayers. J. Chem. Theor. Comput. **11**, 5973–5979 (2015)

7. Vila-Vicosa, D., Teixeira, V.H., Santos, H.A., Baptista, A.M., Machuqueiro, M.: Treatment of ionic strength in biomolecular simulations of charged lipid bilayers. J. Chem. Theor. Comput. **10**, 5483–5492 (2014)

8. Hugot, S., Sy, D., Ruiz, S., Charlier, M., Spotheim-Maurizot, M., Savoye, C.: Radio-protection of dna by spermine: a molecular modelling approach. Int. J. Radiat. Biol. **75**, 953–961 (1999)

9. Gomes, P.J., Ribeiro, P.A., Shaw, D., Mason, N.J., Raposo, M.: UV degradation of deoxyribonucleic acid. Polym. Degrad. Stab. **94**, 2134–2141 (2009)

10. Gomes, P.J., Coelho, M., Dionísio, M., António Ribeiro, P., Raposo, M.: Probing radiation damage by alternated current conductivity as a method to characterize electron hopping conduction in DNA molecules. Appl. Phys. Lett. **101**, 123702 (2012)

11. Gomes, P., et al.: Energy thresholds of DNA damage induced by UV radiation: an XPS study. J. Phys. Chem. B **119**, 5404–5411 (2015)

12. Moraes, M.L., et al.: Polymeric scaffolds for enhanced stability of melanin incor-porated in liposomes. J. Colloid Interface Sci. **350**, 268–274 (2010)

13. Duarte, A., et al.: Characterization of PAH/DPPG layer-by-layer films by VUV spectroscopy. Eur. Phys. J. E Soft Matter **36**, 9912 (2013)

14. Duarte, A.A., et al.: DPPG liposomes adsorbed on polymer cushions: effect of roughness on amount, surface composition and topography. J. Phys. Chem. B **119**, 8544–8552 (2015)

15. Polewski, K., Kniat, S., Slawinska, D.: Gallic acid, a natural antioxidant, in aqueous and micellar environment: spectroscopic studies. Curr. Top. Biophys. **26**, 217–227 (2002)

16. Lin, X.-Q., Li, F., Pang, Y.-Q., Cui, H.: Flow injection analysis of gallic acid with inhibited electrochemiluminescence detection. Anal. Bioanal. Chem. **378**, 2028–2033 (2004)

17. Ari, T., Güven, M.H.: Valence-shell electron energy-loss spectra of formic acid and acetic acid. J. Electron Spectrosc. Relat. Phenom. **106**, 29–35 (2000)

18. Barnes, E.E., Simpson, W.T.: Correlations among electronic transitions for car-bonyl and for carboxyl in the vacuum ultraviolet. J. Chem. Phys. **39**, 670–675 (1963)

19. Xu, K., Amaral, G., Zhang, J.: Photodissociation dynamics of ethanol at 193.3 nm. J. Chem. Phys. **111**, 6271–6282 (1999)
20. Satyapal, S., Park, J., Bersohn, R., Katz, B.: Dissociation of methanol and ethanol activated by a chemical reaction or by light. J. Chem. Phys. **91**, 6873–6879 (1989)
21. Wen, Y., Segall, J., Dulligan, M., Wittig, C.: Photodissociation of methanol at 193.3 nm. J. Chem. Phys. **101**, 5665–5671 (1994)
22. Pawlikowska-Pawlęga, B., et al.: Localization and interaction of genistein with model membranes formed with dipalmitoylphosphatidylcholine (DPPC). Biochimica et Biophysica Acta (BBA)-Biomembranes **1818**, 1785–1793 (2012)
23. Phan, H.T., et al.: Structure-dependent interactions of polyphenols with a biomimetic membrane system. Biochimica et Biophysica Acta (BBA)-Biomembranes **1838**, 2670–2677 (2014)

Inference in a Partial Differential Equations Model of Pulmonary Arterial and Venous Blood Circulation Using Statistical Emulation

Umberto Noè[1(✉)], Weiwei Chen[2], Maurizio Filippone[3], Nicholas Hill[1], and Dirk Husmeier[1]

[1] School of Mathematics and Statistics, University of Glasgow,
Glasgow G12 8SQ, UK
u.noe.1@research.gla.ac.uk
[2] Research Center for Regenerative Medicine,
Guangxi Medical University, Nanning 530021, China
[3] Eurecom, Campus SophiaTech, 450 Route des Chappes, Biot, France

Abstract. The present article addresses the problem of inference in a multiscale computational model of pulmonary arterial and venous blood circulation. The model is a computationally expensive simulator which, given specific parameter values, solves a system of nonlinear partial differential equations and returns predicted pressure and flow values at different locations in the arterial and venous blood vessels. The standard approach in parameter calibration for computer code is to emulate the simulator using a Gaussian Process prior. In the present work, we take a different approach and emulate the objective function itself, i.e. the residual sum of squares between the simulations and the observed data. The Efficient Global Optimization (EGO) algorithm [2] is used to minimize the residual sum of squares. A generalization of the EGO algorithm that can handle hidden constraints is described. We demonstrate that this modified emulator achieves a reduction in the computational costs of inference by two orders of magnitude.

Keywords: Statistical inference · Gaussian processes · Emulation · Simulator · Global optimization · Efficient global optimization · Hidden constraints · Nonlinear differential equations · Pulmonary blood circulation · Pulmonary hypertension

1 Introduction

Patients suffering from chronic pulmonary arterial hypertension (high blood pressure), which is a disease of the small pulmonary arteries, suffer from reduced pulmonary function and right ventricle hypertrophy leading to right heart failure (Rosenkranz and Preston [10]). For diagnosis and ongoing treatment and assessment, clinicians measure blood flow and pressure within the pulmonary arteries. The pressure measurements require right heart catheterization which is invasive,

© Springer International Publishing AG 2017
A. Bracciali et al. (Eds.): CIBB 2016, LNBI 10477, pp. 184–198, 2017.
DOI: 10.1007/978-3-319-67834-4_15

and thus unpleasant for patients and carries risk. Therefore, data about healthy patients are not available due to ethical reasons. The present work uses a partial differential equations (PDEs) model of the pressure and flow wave propagation in the pulmonary circulation under normal physiological and pathological conditions. This model is an extension of previous studies, which considered only the arterial system or part of the venous system [6,8]. The PDEs depend on various physiological parameters, related e.g. to blood vessel geometry, vessel stiffness and fluid dynamics. These parameters can typically not be measured in vivo and hence need to be inferred indirectly from the observed blood flow and pressure distributions. In principle, this is straightforward. Under the assumption of a suitable noise model, the solutions of the PDEs define the likelihood of the data, and the parameters can then be inferred in a maximum likelihood sense. However, a closed-form solution of the maximum likelihood equations is not available, which calls for an iterative optimization procedure. Since a closed-form solution of the PDEs is not available either, each optimization step requires a numerical solution of the PDEs. This is computationally expensive, especially given that the likelihood function is typically multi-modal, and the optimization problem is NP-hard. In the present work, our goal is to reduce the computational costs of inference with the concept of *emulation*. This is to be distinguished from the explicit numerical solution of the PDEs, which we henceforth refer to as *simulation* (or *simulator* when referring to a specific solution). The parameters of the PDEs to be estimated will give clinicians insights into the patient specific vessel structure that would not be obtainable in vivo such as vessel stiffnesses, a primary indication of hypertension.

2   Model

In our model of the pulmonary circulation, seven large arteries and four large veins are modelled explicitly, while the smaller vessels are represented by structured trees (Fig. 1). A magnetic resonance imaging (MRI) based measurement of the right ventricular output provides the inlet flow for the system.

The large arteries and veins are modelled as tapered elastic tubes, and the geometries are based on measurements of proximal and distal radii and vessel lengths [8]. The cross-sectional area averaged blood flow and pressure are predicted from a non-linear model based on the incompressible Navier–Stokes equations for a Newtonian fluid [6].

The small arteries and veins are modelled as structured trees at each end of the terminal large arteries and veins to mimic the dynamics in the vascular beds [8]. With a given parent vessel radius r_p, the daughter vessels are scaled linearly with radii $r_{d_1} = \alpha r_p$ and $r_{d_2} = \beta r_p$, where α and β are the scaling factors. The vessels bifurcate until the radius of each terminal vessel is smaller than a given minimum $r_{\min}$. The radius relation at bifurcations is

$$r_p^\xi = r_{d_1}^\xi + r_{d_2}^\xi, \quad 2.33 \leq \xi \leq 3.0, \tag{1}$$

where the exponent $\xi = 2.33$ corresponds to laminar flow, $\xi = 3.0$ corresponds to turbulent flow [6], p represents the parent vessel, and d_1 and d_2 represent the

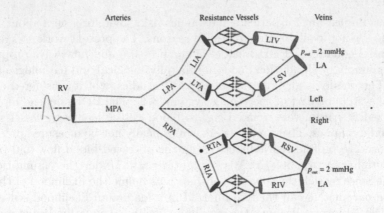

Fig. 1. Schematic of the pulmonary circulation consisting of large arteries, arterioles, venules and large veins from [8]. Seven large arteries are considered in this model, i.e. the main pulmonary artery (MPA), the left (LPA) and right (RPA) pulmonary arteries, the left interlobular artery (LIA), the left trunk artery (LTA), the right interlobular artery (RIA), and the right trunk artery (RTA). The four terminal arteries LIA, LTA, RIA, and RTA are connected to four large veins, i.e. the left inferior vein (LIV), left superior vein (LSV), right inferior vein (RIV), and right superior vein (RSV), via structured trees of resistance vessels.

daughter vessels. Given the area ratio $\eta = (r_{d_1}^2 + r_{d_2}^2)/r_p^2$ and the asymmetry ratio $\gamma = (r_{d_2}/r_{d_1})^2$, the scaling factors α and β satisfy $\alpha = (1 + \gamma^{\xi/2})^{-1/\xi}$ and $\beta = \alpha\sqrt{\gamma}$. The parameters, ξ, γ, $r_{\min}$ and a given root radius r_0, determine the size and density of the structured tree. The cross-sectional area averaged blood flow and pressure in these small arteries and veins are computed from the linearized incompressible axisymmetric Navier–Stokes equations [8].

The system of nonlinear partial differential equations is given by Qureshi et al. [8], and its numerical solution, which depends on various physiological parameters, will henceforth be referred to as the simulator.

Particular interest lies in the estimation of the parameter ξ, because low values are indicative of the clinically relevant problem of vascular rarefaction[1], as in pulmonary hypertension. Its estimation is performed in the range $2.33 \leq \xi \leq 3$, keeping the other parameters of the model fixed to biologically relevant values from the literature [8], and by using pressure and flow measurements from the large vessels' midpoint location to resemble clinical data, corrupted by noise with a signal-to-noise ratio (SNR) of 10 db. Other relevant parameters of interest for clinical diagnosis are the stiffness parameters in the large and small vessels, f_L and f_S respectively. These correspond to Eh/r_0 in Eq. (4) of Qureshi et al. [8], where E denotes Young's modulus, h the vessel wall thickness and r_0 the vessel radius for a given pressure.

---

[1] Vascular rarefaction is an old finding in patients with hypertension, and represents the condition of having fewer blood vessels per tissue volume.

Computational inference of the biophysical parameters entails repeated forward simulations for different parameter configurations. In this model a forward simulation takes around 23 s of CPU time. Each simulation's output is a 22-dimensional vector containing 7 large arteries +4 large veins pressure and flow measurements in the midpoint location of each large vessel. Given the multi-modality of the objective function, a standard global optimization algorithm requires a large number of forward simulations, which comes at substantial computational costs even for the inference of just a single parameter. In the following section, we discuss a faster method based on the concepts of statistical emulation and Bayesian optimization, which aim to optimize an expensive function with the smallest number of function evaluations possible.

3 Method

To perform efficient optimization of a computationally expensive objective function $y(\cdot)$, we let a statistical emulator guide the optimization process. The emulator $f(\cdot)$ of $y(\cdot)$ is based on a Gaussian Process (GP) prior with Matérn 5/2 kernel function[2] k_ψ, as in Snoek et al. [12], depending on hyperparameters ψ inferred in a maximum likelihood sense. In ψ we allow for a different length-scale in each dimension. The Matérn class leads to twicely differentiable sampled paths, which is the same assumption required for example by Quasi-Newton methods like BFGS or LBFGS. Using what has nowadays become the standard covariance function choice in applied statistics, namely the Squared Exponential kernel, would lead to infinitely-differentiable functions, which turns out to be a very unrealistic assumption in many scenarios like time series analysis or more generic engineering and financial applications.

Unlike standard emulation, that models the simulator's output, we aim to directly emulate the objective function $y(\cdot) : \mathbb{X} \to \mathbb{R}$, where $\mathbb{X} \subset \mathbb{R}^d$ is the biophysical parameter space. We use the following hierarchical Bayesian non-parametric regression model:

$$\mathbf{y} \mid \mathbf{f}, \sigma^2 \sim \mathrm{N}(\mathbf{f}, \sigma^2 \mathbf{I}) \tag{2}$$

$$f(\mathbf{x}) \mid m, k \sim \mathrm{GP}(m(\mathbf{x}), k(\mathbf{x}, \mathbf{x}')) \tag{3}$$

where $\mathbf{x}, \mathbf{x}' \in \mathbb{X}$, $\mathbf{y} = [y(\mathbf{x}_1), \dots, y(\mathbf{x}_n)]^T$, $\mathbf{f} = [f(\mathbf{x}_1), \dots, f(\mathbf{x}_n)]^T$ and we assume throughout the analysis that $m(\mathbf{x}) = c$, $\forall \mathbf{x} \in \mathbb{X}$, with c constant to be estimated from the data. This corresponds to the standard literature approach of applying a zero-mean GP to zero-centred data. Given that we are working with black-box functions, which often cannot be visualized because of the high dimensionality or the significant computational complexity, this is a reasonable assumption. If additional knowledge about the functional behaviour for arguments $\mathbf{x}$ far outside the training domain, $|\mathbf{x}| \gg 0$, is available, this can be incorporated into the mean, $m(\mathbf{x})$. However, given that we are operating on a compact rather than open argument set, this is not an issue in this work. The model's hyperparameters are

---

[2] See Eq. (4.14) in Rasmussen and Williams [9] Sect. 4.2.

estimated, as customary in literature, by maximizing the marginal log likelihood of the model. The presence of extreme outliers can be dealt with by using a long-tailed distribution for the observational noise in (2), like a t-distribution. The Student-t likelihood model is included in standard GP software implementations such as GPstuff [15].

The starting point of the optimization should be a good initial picture of the objective function, which is obtained by conditioning the GP on the objective evaluated at a set of design points in the input (i.e. parameter) space. In the present work, we follow Jones et al. [2] and use a space filling Latin Hypercube (LH) design, with the number of initial input points set to $n = 10 \times d$.

To minimize the evaluation-costly objective function we use a sequential strategy proposed in [2], called Efficient Global Optimization (EGO), which selects iteratively the point with the highest expected improvement over the incumbent minimum. Let the random variable *improvement* be $I(\mathbf{x}) = \max\{y_{\min} - f(\mathbf{x}), 0\}$, where

- $f(\mathbf{x}) \sim \mathrm{N}(m(\mathbf{x}), s^2(\mathbf{x}))$ is the marginal GP at the point of interest $\mathbf{x}$;
- $y_{\min} = y(\mathbf{x}_{\min})$ is the best function value known so far;
- $I(\mathbf{x}) > 0$ if $\mathbf{x}$ has a lower function value than the incumbent solution;
- $I(\mathbf{x}) = 0$ otherwise.

The expected improvement (EI) acquisition function (Jones et al. [2]) is the expected value of the random variable $I(\mathbf{x})$ and has the closed-form expression:

$$\mathrm{EI}(\mathbf{x}) = (y_{\min} - m(\mathbf{x}))\Phi\left(\frac{y_{\min} - m(\mathbf{x})}{s(\mathbf{x})}; 0, 1\right) + s(\mathbf{x})\phi\left(\frac{y_{\min} - m(\mathbf{x})}{s(\mathbf{x})}; 0, 1\right), \quad (4)$$

where $\Phi(x; \mu, \sigma^2)$ and $\phi(x; \mu, \sigma^2)$ denote the cumulative distribution function (cdf) and probability density function (pdf) of a $\mathrm{N}(\mu, \sigma^2)$ random variable evaluated at x, respectively. The EI balances *exploitation* (using the predicted objective by the emulator) and *exploration* (improving the emulator where it predicts high uncertainty). This is easily seen by the fact that the contribution of $(y_{\min} - m(\mathbf{x}))$ to (4) is higher when the prediction is smaller than the observed minimum. At the same time $s(\mathbf{x})$ will increase the acquisition value when the GP uncertainty is high at $\mathbf{x}$. The problem of directly optimizing the objective function $y(\cdot)$ derived from the computationally expensive simulator, is now shifted to the maximization of the computationally cheaper EI acquisition function or, equivalently, the minimization of $-\log \mathrm{EI}(\mathbf{x})$. Even if this function can be highly multi-modal, it can be efficiently optimized using multiple restarts or standard state-of-the-art global optimization solvers, like the Dividing Rectangles algorithm [7], as the computational costs for obtaining EI are negligible to those required for computing $y(\mathbf{x})$. Once the minimum $\mathbf{x}_*$ of $-\log \mathrm{EI}(\mathbf{x})$ has been found, we compute the expensive objective function at the next best candidate $\mathbf{x}_*$ and obtain the output $y_* = y(\mathbf{x}_*)$. The new point $(\mathbf{x}_*, y_*)$ is then added to the training dataset $\mathcal{D} = \{\mathbf{X}, \mathbf{y}\} \cup \{\mathbf{x}_*, y_*\}$ and a new iteration starts by re-fitting the GP on the augmented data $\mathcal{D}$. The process continues iteratively until convergence or until the maximum budget of function evaluations has been exceeded. The point $(\mathbf{x}_i, y_i) \in \mathcal{D}$ having minimum observed objective $y_i = y(\mathbf{x}_i)$ is the returned EGO minimizer of $y(\cdot)$.

4   Illustration

We now illustrate the EGO algorithm on the 'Sasena' function [11], shown in Fig. 2(a), where at each iteration we sample the objective at the input that minimizes $-\log \mathrm{EI}$:

$$y(\mathbf{x}) = 2 + 0.01(x_2 - x_1^2)^2 + (1 - x_1)^2 + 2(2 - x_2)^2 + 7\sin(0.5x_1)\sin(0.7x_1x_2)$$
$$0 \le x_1 \le 5, \ 0 \le x_2 \le 5. \tag{5}$$

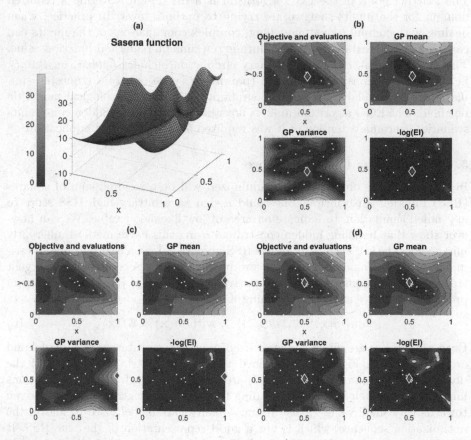

Fig. 2. EGO algorithm with Matérn $\nu = 5/2$ kernel. (a) Sasena function (5), scaled to $[0,1]^2$, to be minimized. (b) EGO 1st iteration and training inputs (white dots). (c) 5th iteration. (d) 25th and last iteration. The red and white diamond represents the next evaluation point, i.e. the minimizer of $-\log \mathrm{EI}$. (Color figure online)

During the first iteration the function is evaluated at the $n = 10 \times 2$ LH design points, white dots in Fig. 2(b,c,d), and training data $\mathcal{D} = \{\mathbf{x}_i, y_i\}_{i=1}^n$ are thus obtained. The GP is trained on the data. We can see that at the beginning, with this particular initial design and chosen number of design points, we already have a fairly good picture of the objective function. Then iteratively the acquisition function is optimized (the diamond represents the optimum), the new sampled point is added to the dataset and the GP is trained again. From the plotted optimizer of the acquisition function, we see that the EI function can be increased when the predicted mean is smaller than the incumbent minimum, Fig. 2(b) and (d), or when the variance is increased, Fig. 2(c).

5   EGO with Hidden Constraints

The starting point of the EGO algorithm is a LH design, covering a required domain for the inputs that we are trying to optimize over. In practice, when dealing with computational modelling, complex combinations of the inputs can cause numerical instabilities and failure in obtaining a requested function value. The PDEs simulator of the pulmonary circulation includes *hidden constraints*, i.e. unknown regions of the domain (parameter space) where a requested simulation is not available because the assumptions of the physiological model do not hold. The EGO algorithm in [2] is not able to handle the hidden constraints scenario, hence an extension of it was required for the presented application.

5.1   Assigning a High RSS Score

In this study, the objective to be minimized will be a residual sum of squares (RSS) function. In principle, one could assign an arbitrary high RSS score to any failed simulation, to indicate an area of low likelihood value. We can however show that handling hidden constraints using this naive method inherently misspecifies the GP surrogate[3] of the RSS objective. Unrealistic and biased estimates of the parameters would be produced, as well as completely divergent traces of the minimum observed objective, $\min\{y_1, \ldots, y_n\}$, and the minimum predicted objective among the training inputs,

$$\min\{m(\mathbf{x}_1), \ldots, m(\mathbf{x}_n)\}, \quad \text{with } m(\mathbf{x}) = \mathbb{E}f(\mathbf{x}). \tag{6}$$

Both pathologies are direct consequence of a non-representative metamodel and are illustrated in Fig. 3. The flattened area of the GP in panel (a) and the divergent minimum traces in panel (d) are due to the high-frequency oscillations induced by the high RSS values, leading to the kernel length-scale being driven towards very small values. This produces a surrogate model, used to inform the optimization sequence, which is not a good representation of the true RSS. It also leads to estimates from the true RSS and the surrogate that are completely contradictory. Compare panel (d) to the traces in Fig. 5(f), obtained from a good model of the objective.

---

[3] In this context, surrogate or metamodel are synonyms for GP posterior mean, and were generated by the engineering community.

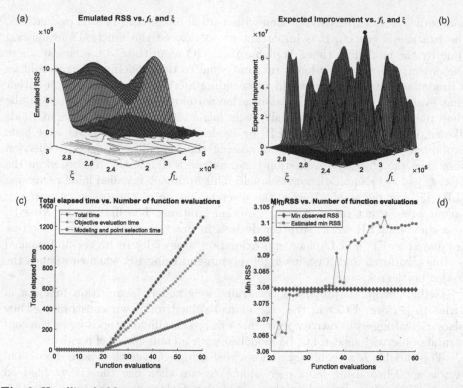

Fig. 3. Handling hidden constraints by assigning a high RSS value and using the EGO algorithm. (a) Surrogate of the RSS objective and training points (dots). (b) EI acquisition function and its maximum (black dot). (c) Total elapsed time trace (diamonds) decomposed as objective evaluation time (squares) plus objective modelling and acquisition maximization time (dots). (d) Traces of the minimum observed RSS (diamonds) and estimated RSS minimum among the visited points (dots), see Eq. (6).

5.2   Constraint Weighted Acquisition Functions

Hidden constraints are common in mathematical modelling, engineering and life sciences applications that include equipment, time, money constraints or partial knowledge of a system. In the latter scenario, failure to obtain a requested output due to a model breakdown is a symptom of incomplete knowledge of the real-world system that we are trying to model. In order to generalize the EGO algorithm to account for hidden constraints, we keep track of where the requested outputs are not available. We define an auxiliary variable $h(\mathbf{x})$ equal to $+1$ in case of a failed simulation at $\mathbf{x}$ and -1 otherwise. The next step is to build a model of failure in obtaining a requested function value. The natural approach would be to use a GP classifier with a probit or logit link function. However, we recall that the strength of the EGO algorithm lies in shifting the optimization of an expensive-to-evaluate objective to the optimization of a cheap acquisition function like the EI. In these settings, where the aim is to minimize a computationally expensive function with a small budget of function evaluations, the runtime is crucial. It is important to not exceed the function's evaluation time by the modelling time. Applying the EGO algorithm to a scenario where the objective is faster, or has a runtime equal to the modelling time, would be a fundamental error as we would be spending more time in modelling it rather than evaluating it. For the GP classification with logit or probit link function, the class posterior probability is analytically intractable. Hence, iterative methods like Expectation Propagation (EP) or Laplace approximation (LA) have been proposed. To avoid significantly increasing the modelling and point selection time over the objective evaluation time, we use a simple GP regression on the $h(\mathbf{x}) \in \{+1, -1\}$ simulation error labels. This approach is called label regression (LR) [5] and purposely ignores the discrete nature of the binary variable in favour of an exact inference and a quicker runtime. The findings by Kuss [3] also show that LR works surprisingly well in practice, with a lower error rate compared to EP and Laplace approximation, especially in higher-dimensional settings. Nickisch and Rasmussen [5] recommend using LR when runtime is the major bottleneck.

Gelbart et al. [1] propose a constraint weighted EI acquisition function in order to perform EGO in the case of real-valued unknown constraints. They also briefly suggest iterative approximations or sampling methods when the constraints observations can not be modelled with a Gaussian likelihood.

The variable $h(\mathbf{x})$ is binary and not real-valued. Given the above discussion, it was modelled using label regression. We can then use the GP trained on failure labels $\mathcal{D}_h = \{\mathbf{x}_i, h_i\}$ to obtain a hidden-constraints-weighted acquisition function.

Given the simulation failure GP regression model with marginals $f_h(\mathbf{x}) \sim N(m_h(\mathbf{x}), s_h^2(\mathbf{x}))$, we can calculate the probability of a successful simulation in any region of the domain as follows:

$$P(h(\mathbf{x}) = -1) = P(y(\mathbf{x}) \text{ successful}) = \Phi(0; m_h(\mathbf{x}), s_h^2(\mathbf{x})), \qquad (7)$$

where $m_h(\mathbf{x}), s_h^2(\mathbf{x})$ are the predictive mean and variance of the failure GP model f_h at the test location $\mathbf{x} \in \mathbb{X}$. As errors are labelled as $+1$ and successful evaluations as -1, by taking the probability of the GP being less than 0 we obtain an indication of the probability of a successful simulation. The use of the Normal cdf follows from the property of Gaussian processes that every finite dimensional distribution is a multivariate Gaussian. Similarly to Gelbart [1], given the probability of a successful simulation, it is possible to obtain a corrected version of any acquisition function $a(\mathbf{x})$ that avoids infeasible regions by weighting $a(\mathbf{x})$ with the probability that the input will lead to a successful simulation. The hidden-constraints-weighted (HCW) acquisition function is then given by the following formula:

$$a_{\text{HCW}}(\mathbf{x}) = a(\mathbf{x}) \times P(h(\mathbf{x}) = -1) = a(\mathbf{x}) \times \Phi(0; m_h(\mathbf{x}), s_h^2(\mathbf{x})). \qquad (8)$$

It is then optimized as described in Sect. 3 and the optimum $\mathbf{x}_\star$ is chosen as the next evaluation point. The algorithm continues by evaluating $y(\mathbf{x}_\star)$ at the best

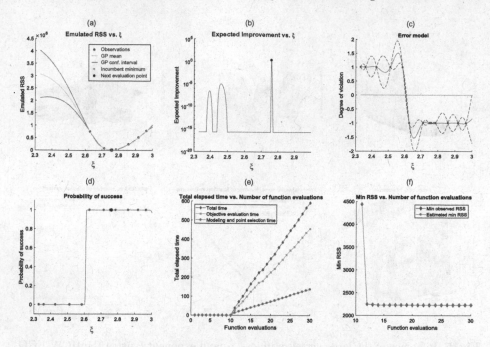

Fig. 4. Inference of the exponent ξ using the HCW-EGO algorithm. (a) Surrogate RSS function. (b) EI_{HCW} acquisition function and its maximizer (dot). (c) GP model of objective evaluation failure. (d) Probability of a successful run of the simulator. (e) Cumulative elapsed time (diamonds) decomposed as simulation time (squares) plus objective modelling time and acquisition maximization time (dots). (f) Traces of the minimum observed RSS (diamonds) and estimated RSS minimum among the visited points (dots), i.e. the minimum emulated objective, among those points at which the objective function was evaluated, see Eq. (6).

candidate $\mathbf{x}_\star$ and by keeping track of the success or failure in the simulation in $h(\mathbf{x}_\star)$. The modified EGO algorithm iteratively maintains both a GP surrogate $f(\cdot)$ of the objective y and a GP regression model $f_h(\cdot)$ of the error labels h. We will call this algorithm the HCW-EGO algorithm and it consists in maintaining two GPs at each iteration: one for the objective emulation, and one for the error labels regression. These two models are iteratively required to evaluate the hidden-constraints-weighted acquisition function. In particular, the acquisition function chosen in this study is the hidden-constraints-weighted EI:

$$\text{EI}_{\text{HCW}}(\mathbf{x}) = \text{EI}(\mathbf{x}) \times P(h(\mathbf{x}) = -1) = \text{EI}(\mathbf{x}) \times \Phi(0; m_h(\mathbf{x}), s_h^2(\mathbf{x})). \quad (9)$$

6   Simulations

In order to assess the proposed inference scheme for the pulmonary circulation model, we simulated pressure and flow data $\mathbf{y}_0 \in \mathbb{R}^{d=22}$ from the PDE model

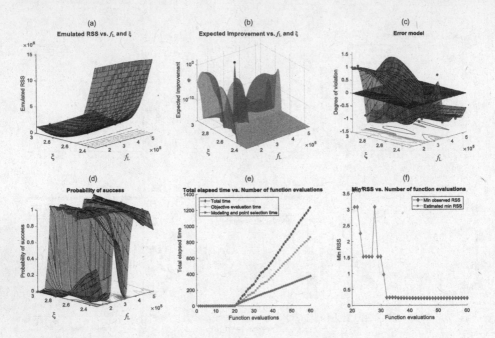

Fig. 5. Inference of the large vessels stiffness f_L and exponent ξ using the HCW-EGO algorithm. (a) Surrogate RSS function. (b) EI_{HCW} acquisition function and its maximizer (dot). (c) GP model of objective evaluation failure. (d) Probability of a successful run of the simulator. (e) Cumulative elapsed time (diamonds) decomposed as simulation time (squares) plus objective modelling time and acquisition maximization time (dots). (f) Traces of the minimum observed RSS (diamonds) and minimum predicted objective among the training inputs (dots), see Eq. (6).

for a given parameter vector $\mathbf{x}_0$ assumed to be the underlying truth, and then added noise with a SNR of 10 db. The observations vector includes pressure and flow measurements from the 11 large vessels' midpoint locations, in order to resemble the data which clinicians would provide. Now, pretending that the true parameter $\mathbf{x}_0$ is unknown, interest lies in its estimation from the noisy observations $\mathbf{y}_0$.

We created a space filling LH design of $n = 10 \times d$ points for $\mathbf{x} \in [\mathbf{x}_L, \mathbf{x}_U] \subset \mathbb{R}^d$, and ran a forward simulation for each of the design points, obtaining a set of simulations $\{\mathbf{x}_i, \mathbf{y}_i\}_{i=1}^n$ and the corresponding success/failure labels $\mathbf{h} = [h_1, \ldots, h_n]^T$.

The chosen objective function is the residual sum of squares (RSS) between the simulations and observations:

$$\mathrm{rss}_i = \|\mathbf{y}_i - \mathbf{y}_0\|^2, \tag{10}$$

and we fitted simultaneously two GPs. The first is the RSS surrogate and is conditioned on the training data $\mathcal{D} = \{\mathbf{X}, \mathbf{rss}\}$, where $\mathbf{rss} = [\mathrm{rss}_1, \ldots, \mathrm{rss}_n]^T$, while the second one is the error GP model, trained on the inputs and error labels recorded during simulation time $\mathcal{D}_h = \{\mathbf{X}, \mathbf{h}\}$. We then applied the modified EGO algorithm using the hidden-constraints-weighted EI acquisition function (9), with input $\mathbf{x}_i$ containing the PDE parameters and output $y_i = \mathsf{rss}_i$.

The first experiment is a one-dimensional inference problem. We tried to estimate $\xi \in [2.33, 3]$, with underlying truth being 2.76, using a budget of 30 function evaluations only. Results for one run of the algorithm, i.e. for a particular design instantiation, are shown in Fig. 4. Next, we focused on a two-dimensional problem, trying to estimate $\mathbf{x} = (f_L, \xi)$ simultaneously, with $f_L \in [1.33 \times 10^5, 5.33 \times 10^5]$. In this scenario the underlying truth was $\mathbf{x}_0 = (2.6 \times 10^5, 2.76)$, and we allowed for a budget of 60 function evaluations. Two-dimensional results for a given LH design are shown in Fig. 5. We then inferred all three parameters of clinical interest simultaneously, $\mathbf{x} = (f_L, f_S, \xi)$, with $f_S \in [2.66 \times 10^4, 1.066 \times 10^5]$. The underlying truth was $\mathbf{x}_0 = (2.6 \times 10^5, 5 \times 10^4, 2.76)$, and we allowed for a budget of 60 function evaluations. The three-dimensional inference results for one particular run of the algorithm are shown in Fig. 6.

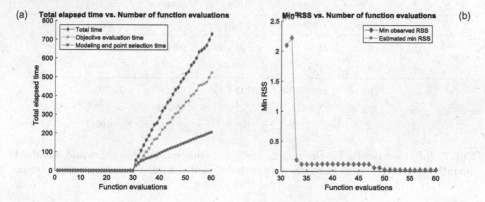

Fig. 6. Inference of the large vessels stiffness f_L, small vessels stiffness f_S and exponent ξ using the HCW-EGO algorithm. (a) Cumulative elapsed time (diamonds) decomposed as simulation time (squares) plus objective modelling time and acquisition maximization time (dots). (b) Trace of the minimum observed RSS (diamonds) and minimum predicted RSS among training inputs (dots), Eq. (6).

7 Results

Figure 7 reports a summary of the results[4] of the inference study, calculated over five independent design instantiations having different random number generator seeds. The first column shows the problem dimensionality, while the second column shows the parameters that have been inferred simultaneously. The 3rd

---

[4] The presented results were obtained on a CentOS 7 machine using MATLAB®. Our code used to perform inference depends on the GPML toolbox by Rasmussen [9] and the standard MATLAB® Statistics Toolbox.

column contains the underlying truth for the parameters. In the fourth column we find the average of the parameters estimated by iterative minimization of the RSS using the EGO algorithm, and their standard deviations over the five runs. Assuming that the final surrogate of the objective is a good model of it, we can in principle improve in the EGO estimated minimum by taking the point that minimizes the surrogate model's posterior mean. These estimates are shown in the fifth column, where the mean and the standard deviation are taken over the five runs for different LH designs. We claim that checking consistency of the 4th and 5th column in each experiment run is a good convergence diagnostics as well as an important measure of the agreement between the surrogate objective model and the true objective. The final column reports the total number of RSS evaluations allowed, i.e. the a priori budget of function evaluations and time allocated to the numerical experiment.

Dim	Parameters	Truth	EGO results		Minimum of RSS surrogate		Iterations budget
d	$\mathbf{x}$	$\mathbf{x}_0$	Mean	St. Dev.	Mean	St. Dev.	
1	ξ	2.76	2.7601	0.0001	2.7601	0.0004	30
2	f_L	2.6×10^5	2.5989×10^5	0.0022×10^5	2.5981×10^5	0.0035×10^5	60
	ξ	2.76	2.76	< 0.0001	2.7604	< 0.0001	
3	f_L	2.6×10^5	2.6178×10^5	0.0212×10^5	2.6101×10^5	0.0091×10^5	60
	f_S	5×10^4	49737	175	50018	67	
	ξ	2.76	2.7597	0.0036	2.7599	0.0013	

Fig. 7. Inference results for the pulmonary circulation model. Averages and standard deviations over five design instantiations with different random number generator seeds.

For the 1D inference problem we find a consistently good estimate by both the observed RSS and the minimum surrogate point, with a very high confidence in the inferred value. Figure 4(a) shows how, by handling the errors correctly, the RSS function has a quadratic-like shape. Panel (b) shows the EI_{HCW} acquisition function. Panel (c) shows the simulation failure GP model that we use to derive the probability of a successful simulation in (d) according to Eq. (7) introduced above. Panel (e) shows the decomposition of the cumulative elapsed time as objective evaluation time plus objective modelling and point selection time. We see that the algorithm is spending more time in evaluating the objective than emulating it and selecting the next best input, so we are gaining computational time by minimizing the number of function evaluations. Panel (f) shows the observed incumbent minimum trace vs the minimum predicted objective among the training inputs, Eq. (6). We see a fast convergence in less than 20 function evaluations, where the value is not changing for the next iterations, but just improving the last decimals. Similar considerations can be done for Figs. 5 and 6.

8   Conclusion

Our aim was to perform inference in a computationally expensive and novel model of the combined arterial and venous pulmonary blood circulation. The parameters of interest are f_L, f_S and ξ. The exponent ξ governs the vessel parent-to-daughter radius relation (1), with low values indicating vascular problems of clinical interest. As ξ increases, the number of vessels in the structured tree will also increase; similarly, as it decreases, the number of vessels will also decrease, simulating the vascular rarefaction clinical condition. The stiffness parameters in large, f_L, and small vessels, f_S, are also of particular interest because stiffening of these vessels is a primary cause of pulmonary arterial hypertension which leads to right heart failure.

In previous studies with state-of-the-art non-emulation-based global optimization algorithms, like Genetic Algorithms or Scatter Search methods [14], we found that the number of required function evaluations was in the order of 2×10^3 for 1D problems, and reached even 10^4 function evaluations for simple 2D or 3D scenarios. Given that the computational costs of a single forward simulation are about 23 s of CPU time, the total computational costs would be in the order of 13 h for 1D inference tasks and could reach two and a half days for 3D inferential problems. The results in Figs. 4, 5(f) and 6(b), show that the proposed emulation-based approach achieves a substantial reduction in the number of forward simulations, effectively converging to a very good estimate of the parameters in less than the allowed function evaluations budget (last column in Fig. 7), while spending the remaining iterations refining the last decimals. In the 1D scenario we reached convergence in less than 11 min, while for the 2D and 3D scenario in about 23 min or less. This corresponds to a total reduction of the computational complexity by two orders of magnitude.

In future work, we will extend the inference to all model parameters, using informative priors to incorporate knowledge from the biophysical literature. However, the purpose of the present proof-of-concept study is to keep the dimension of the parameter space low, so as to allow a visualization of the results (see Figs. 4 and 5). We will also apply the proposed emulation scheme to the Holzapfel–Ogden soft tissue mechanics model of the heart [16], where the computational costs of a single forward simulation are in the order of an hour, and the overall computational savings promise to be even more substantial.

Acknowledgment. UN is supported by a scholarship from the Biometrika Trust. SofTMech is a research centre for Multi-scale Modelling in Soft Tissue Mechanics, funded by EPSRC (grant no. EP/N014642/1).

References

1. Gelbart, M.A., Snoek, J., Adams, R.P.: Bayesian optimization with unknown constraints. In: Uncertainty in Artificial Intelligence (UAI) (2014)
2. Jones, D.R., Schonlau, M., Welch, W.J.: Efficient global optimization of expensive black-box functions. J. Global Optim. **13**, 455–492 (1998)

3. Kuss, M.: Gaussian process models for robust regression, classification, and reinforcement learning. Ph.D. thesis, Technische Universität, Darmstadt (2006)
4. Mockus, J., Tiesis, V., Zilinskas, A.: The application of bayesian methods for seeking the extremum. Towards Global Optim. **2**, 117–129 (1978)
5. Nickisch, H., Rasmussen, C.E.: Approximations for binary gaussian process classification. J. Mach. Learn. Res. **9**, 2035–2078 (2008)
6. Olufsen, M.S.: Structured tree outflow condition for blood flow in larger systemic arteries. Am. J. Physiol. Heart Circulatory Physiol. **276**, 257–268 (1999)
7. Perttunen, C.D., Jones, D.R., Stuckman, B.E.: Lipschitzian optimization without the lipschitz constant. J. Optim. Theory Appl. **79**(1), 157–181 (1993)
8. Qureshi, M.U., Vaughan, G.D.A., Sainsbury, C., et al.: Numerical simulation of blood flow and pressure drop in the pulmonary arterial and venous circulation. Biomech. Model. Mechanobiol. **13**(5), 1137–1154 (2014)
9. Rasmussen, C.E., Williams, C.K.I.: Gaussian Processes for Machine Learning. The MIT Press, Cambridge (2005)
10. Rosenkranz, S., Preston, I.R.: Right heart catheterisation: best practice and pitfalls in pulmonary hypertension. Eur. Respir. Rev. **24**, 642–652 (2015)
11. Sasena, M.J.: Optimization of Computer Simulations via Smoothing Splines and Kriging Metamodels. MSc Thesis, University of Michigan (1998)
12. Snoek, J., Larochelle, H., Adams, R.P.: Practical bayesian optimization of machine learning algorithms. In: Advances in Neural Information Processing Systems, pp. 2951–2959 (2012)
13. Snoek, J.: Bayesian Optimization and Semiparametric Models with Applications to Assistive Technology. Ph.D. thesis, University of Toronto, Toronto, Canada (2013)
14. Ugray, Z., et al.: Scatter search and local nlp solvers: a multistart framework for global optimization. INFORMS J. Comput. **19**(3), 328–340 (2007)
15. Vanhatalo, J., et al.: GPstuff: Bayesian modeling with Gaussian processes. J. Mach. Learn. Res. **14**(1), 1175–1179 (2013)
16. Wang, H.M., et al.: Structure-based finite strain modelling of the human left ventricle in diastole. Int. J. Numer. Methods Biomed. Eng. **29**, 83–103 (2013)

Ensemble Approaches for Stable Assessment of Clusters in Microbiome Samples

Sanja Brdar$^{(\boxtimes)}$ and Vladimir Crnojević

BioSense Institute, University of Novi Sad,
Dr Zorana Djindjica 1, Novi Sad, Serbia
{brdars,crnojevic}@uns.ac.rs
http://www.biosens.rs

Abstract. Fundamental endeavour to understand microbiome and its functions starts with detecting which microbes are present in the samples and continues with comparing different samples and finding similar based on their community compositions. Pervasive method to accomplish these steps is clustering. However clustering brings number of possibilities regarding algorithms, parameters, distance/similarity metrics, etc., that produce different outcomes making it hard to interpret results. The study presented here examines the stability of clusters in the context of various beta diversity metrics applied on human microbiome samples. We explored the effects of 24 different diversity metrics on clustering outcomes and their impact on the accuracy of the clustering of microbiome samples. To overcome obscure results coming from individual clusterings that rely on distinct beta diversity metrics we employed two ensemble approaches to integrate results of individual clusterings. Obtained results on human microbiome data imply that ensemble clustering approaches produce stable results in reconstructing clusters that correspond to the different host and body habitat.

Keywords: Ensemble clustering · Metabarcoding · Microbial communities · Diversity

1 Introduction

High-throughput experiments revolutionize microbial ecology by increasing the speed of research and discoveries related to the diversity of microorganisms and their roles in ecological processes [1]. Current studies investigate microbial communities extracted directly from the environment and sequenced with NGS technology. They aim at understanding microorganisms that exists in different environments: human (gut, skin, oral...), water, soil. The question: "Who is there?" comes first in studying microbiome sample. Identifying which microbes are present and quantifying their abundances provides insights into the diversity of the examined sample. Currently prevailing technique in studying microbial diversity is sequencing of marker genes 16S (prokaryotic) or 18S (eukaryotic) rRNA, that are highly conserved between different species and thus suitable for

© Springer International Publishing AG 2017
A. Bracciali et al. (Eds.): CIBB 2016, LNBI 10477, pp. 199–208, 2017.
DOI: 10.1007/978-3-319-67834-4_16

phylogenetic taxonomy. Such approaches are denoted as DNA metabarcoding [2] and characterized as economic way of taxonomic identification that enables monitoring diversity and comparisons of taxonomic compositions among various environmental samples.

Taxonomy relies on clustering analysis i.e. grouping similar species into clusters. Groups of microbial species that show a certain level of similarity represent operational taxonomic units (OTUs). After identifying OTUs in a multiple samples under the analysis, the next step includes between-sample comparisons based on some distance measure, termed as beta diversity analysis and then again applying clustering to identify communities among samples. But clustering brings numerous users' dilemmas such as selecting algorithm, parameters, similarity/distance metrics, thresholds, etc.

Although an importance of studying complex microbial communities in a natural environments is recognized, studies that address reliability of derived conclusions are just recently increasingly appreciated. Inconsistent results may be implication of unstable OTUs obtained by *de novo* clustering [3], different diversity measures [4] or as examined in the detection of enterotypes, results may be affected by OTU taxonomic level, OTU-picking method, 16S rRNA variable region and most substantially by distance metric and the clustering score method [5].

Ensemble clustering approaches hold potential for improving the robustness, stability and accuracy of discovered clusters. In this light, microbial diversity analysis may also benefit from an integration across multiple partitions. Benefits of the integration across different clusterings algorithms and parameters have been recently evidenced [6]. Here, in our study, integration scenario covers different beta diversity measures [7]. We used 24 beta diversity measures to quantify pairwise differences among samples and then ran spectral clustering [8] on the similarity matrices obtained by transforming pairwise distances. Finally, for the assessment of communities in microbiome samples we integrated the results of individual clusterings and applied two ensemble approaches–one recently proposed that utilizes non-negative matrix factorization - NMF [9] and another well known consensus clustering - CONS [10].

2   Materials and Methods

2.1   Data

We used data from "Moving pictures of the human microbiome" study [11]. Data set encompasses approximately 69 million sequences obtained form NGS (*next-generation sequencing*) experiment that included 1967 microbiome samples extracted from oral, skin and gut sites on the human body of two individuals, female and male, sampled over 396 time points. Differences in microbial compositions between body sites and individuals were relatively stable over time what makes data set suitable for evaluating clustering algorithms. Data were accessed through MG-RAST API [12] after quality filtering step. Overall size of the set is ≈ 12 GB. Sample labels that indicate microbiome host and body site

Table 1. Experimental data.

Gender	Place	Number of samples
Female	Oral	135
Female	Skin	268
Female	Gut	131
Male	Oral	373
Male	Skin	724
Male	Gut	336

were extracted from corresponding metadata. Number of samples across labels defined by gender/body sites are presented in Table 1.

2.2 Methods

To perform microbial community analysis we used QIIME package [13] extended with individual and ensemble clustering algorithms. QIIME package includes a large number of tools for processing and analysing microbial sequence data. Various pipelines can be performed starting from the raw sequence data to the final diversity analysis and visualizations. The steps that were conducted in our experiments include:

1. OTU picking
2. Making OTU biom table
3. Measuring beta diversity among samples
4. Spectral clustering
5. Ensemble clustering

Diversity studies can be reference-based, i.e. rely on sequence similarity against reference database or reference-free where sequences are clustered based on the similarities to one another. In the first approach clustering can be performed largely in parallel, but only sequences that match a sequence in a reference database with high similarity are clustered while those below defined threshold are discarded. In the reference-free clustering, refereed as *de-novo*, all reads are clustered, but the process is not easily parallelized. Recently proposed subsampled open-reference OTU picking method [14] provides trade-of between these two. Here, we used reference-based approach to produce stable OTUs. Sequences were clustered into OTUs by default taxonomy assigner - UCLUST [15] with a sequence similarity threshold of 97% or 99% against Greengenes reference database [16]. Threshold of 97% is a commonly used rule of thumb to define species, but also tighter threshold of 99% have been proposed. Therefore, we explored both. Clustering algorithm, UCLUST is a greedy algorithm. Given the query sequence, it searches database of reference sequences. If UCLUST finds a sequence in the reference collection with similarity greater than or equal to defined threshold, it creates OTU defined by the reference sequence and assigns query sequence to it, otherwise query sequence is discarded. The result of clustering sequences is OTU table that summarizes taxonomy of samples in a form of observations counts per-sample.

To quantify beta diversity, that is a diversity between samples, we explored 24 non-phylogenetic beta diversity measures: (1) abundance weighted Jaccard distance, (2) binary Chi-square, (3) binary Chord, (4) binary Euclidean distance, (5) binary Hamming (6) binary Jaccard (7) binary Lennon (8) binary Ochiai (9) binary Pearson, (10) binary Srensen-Dice (11) Bray-Curtis (12) Canberra, (13) Chi-square, (14) Chord, (15) Euclidean, (16) Gower, (17) Hellinger, (18) Kulczynski, (19) Manhattan distance, (20) Morisita-Horn, (21) Pearson, (22) Soergel, (23) Spearman rank and (24) Species profile distance.

Previously described steps form QIIME workflow were extended with two additional: one that runs spectral clustering and the other that integrates clustering results and creates final clusters. Spectral clustering was selected due to its property that can work directly with pair-wise distances/similarities. Pairwise differences among samples were transformed into similarities by using element-wise transformation: $S = e^{-D/(2\mu^2)}$, where D is a pair-wise beta diversity matrix, μ is mean value of that matrix, and S is the final similarity matrix. Spectral clustering then uses S as input.

The results of individual clustering on different pairwise distance matrices are combined to perform ensemble clustering. In NMF approach ensemble is represented as matrix of cluster memberships $R = \{0,1\}^{m \times n}$, where one dimension represents clusters (m is the total number of clusters produced by individual clusterings) and the other samples (n is the total number of examined samples). NMF finds an approximation $R \approx WH$, where W and H are two non-negative factors such that $W \in \mathbb{R}^{m \times k}$ and $H \in \mathbb{R}^{k \times n}$. Parameter k is a factorization rank and equals to the target number of clusters. In the resulting factorization the matrix W contains encoding coefficients while rows of H are the basis vectors that can be interpreted as continuous memberships to target clusters discovered by factorization. Other approach, consensus clustering, integrates cluster memberships into a consensus matrix $\mathbb{R}^{n \times n}$, where indices correspond to samples. In a pairwise manner, matrix sums number of times each two samples were clustered together and divides it by number of times they were both present in the clustering. Final values range between 0 and 1. Consensus matrix can be viewed as a similarity matrix and post-processed through additional clustering methods to obtain final clusters. Here we used agglomerative hierarchical clustering on consensus matrix to produce final clusters.

3   Results

3.1   Individual and Ensemble Clustering Evaluation

To evaluate effectiveness of the clusterings on microbiome samples we employed V-measure, that is a harmonic mean between homogeneity and completeness [17], and adjusted rand index [18], two commonly used measures for evaluating clusters against true labels [19]. Here we measured how clustering results align with 6 labels corresponding to different gender/body sites (see Table 1). Examined microbiome samples belong to a time series study. Although temporal variation exists, stable patterns among body habitats and individuals emerge, thus making the data suitable for benchmarking of clustering algorithms. We evaluate all individual clusterings obtained with spectral clustering algorithm on different beta diversity matrices and two ensemble approaches. The results for performed experiments, where similarity cut-off was set to 97%, are summarized in Figs. 1 and 2.

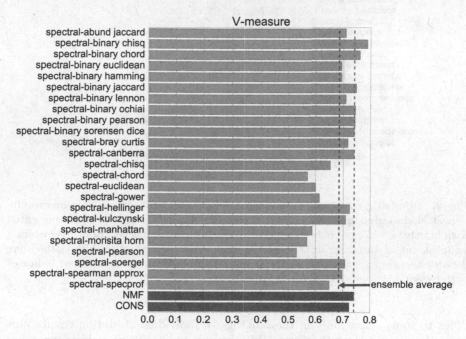

Fig. 1. V-measure between cluster labels and true labels. Baseline results - spectral clustering applied on different pairwise diversity matrices - are represented with blue bars, while red bars correspond to integrative approaches. Prior to clustering samples, cut-off threshold in OTU-picking was set to 97%. Vertical blue dashed line denotes average V-measure of the ensemble's ingredients and red indicates better ensemble approach. (Color figure online)

Results of spectral clusterings on different beta diversity measures and integrative clusterings by NMF and CONS are presented with horizontal bars. Vertical blue dashed line denotes average performance of assembling partitions, and red dashed line highlights the score of better ensemble approach. Figures unveil that ensemble clusterings outperform an average performance of individual clusterings, NMF reached better result than CONS, and it was slightly below the best individual clustering score in the ensemble. We can observe variability of the obtained results elicited by chosen distance measure. If we compare results by the used evaluation measures, adjusted rand index or V-measure, the results

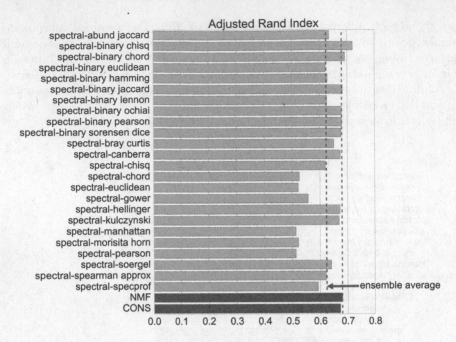

Fig. 2. Adjusted rand index between cluster labels and true labels. Baseline results - spectral clustering applied on different pairwise diversity matrices - are represented with blue bars, while red bars correspond to integrative approaches. Prior to clustering samples, cut-off threshold in OTU-picking was set to 97%. Vertical blue dashed line denotes average adjusted rand index of the ensemble's ingredients and red indicates better ensemble approach. (Color figure online)

differ to some extent only in the rankings of individual clustering results, but general conclusions are the same. NMF, as well as CONS, ensemble approaches provided result that overcomes dependencies on underlying diversity measures. This results is confirmed by both evaluation measures.

Similar results were obtained on different sequence similarity cut-off of 99% (Figs. 3 and 4). NMF and CONS, outperformed average score of individual clustering that entered ensemble. Scores among individual clusterings changed, as well as their rankings. The best score came from other diversity measure, while ensemble clusterings remained stable.

Although results of spectral clustering with some beta diversity measures slightly surpass ensemble approaches, difficulty arise from the selection of appropriate measure for particular data set and other analysis settings. The best V-measure score among individual clusterings in experiments with cut-off of 97%, was produced on beta diversity matrix measured by binary Chi-square (Fig. 1) and on 99% cut-off (Fig. 3) the best score comes from Kulczynski measure. For adjusted rand index score, the best result among individual clusterings for 97% cut-of was obtained with binary Chi-square and for 99% cut-off with Chi-square. Related to Chi-square measure, we can observe how small change

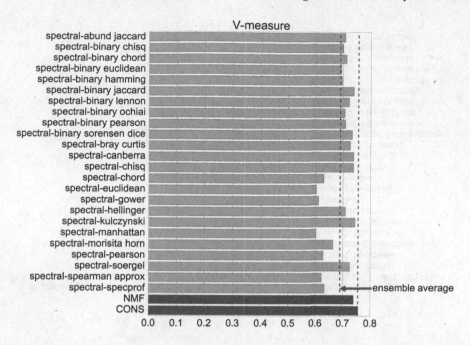

Fig. 3. V-measure between cluster labels and true labels. Baseline results - spectral clustering applied on different pairwise diversity matrices - are represented with blue bars, while red bars correspond to integrative approaches. Prior to clustering samples, cut-off threshold in OTU-picking was set to 99%. Vertical blue dashed line denotes average V-measure of the ensemble's ingredients and red indicates better ensemble approach. (Color figure online)

in cut-off threshold highly impacts outcome from the best to below average of the ensemble. Interestingly, Gower and Canberra distances, recommended as the well performing for detecting clusters [4], here produced divergent results. While Canberra distance was among better metrics, but still below NMF ensemble, Gower distance failed to detect clusters that align with labels of human microbiome data set.

3.2 Stability Analysis

To further examine stability of results, we performed random selection of 1000 out of 1967 samples. This subsampling and overall process of OTU-picking, forming OTU tables, calculating beta diversities, clustering and assembling was repeated 50 times. The results are summarized by box plots (Fig. 5), one for each of the approaches - spectral clustering combined with different diversity measures, NMF and CONS ensemble clustering. We can observe high variability of individual clusterings and improved stability of the ensemble clusterings. Running experiments on subsamples allowed us to measure statistical significance. ANOVA tests indicate that significant difference exists among different methods

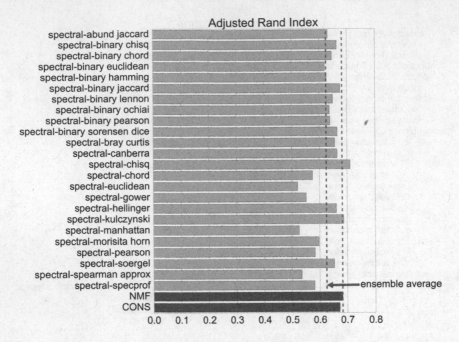

Fig. 4. Adjusted rand index between cluster labels and true labels. Baseline results - spectral clustering applied on different pairwise diversity matrices - are represented with blue bars, while red bars correspond to integrative approaches. Prior to clustering samples, cut-off threshold in OTU-picking was set to 99%. Vertical blue dashed line denotes average adjusted rand index of the ensemble's ingredients and red indicates better ensemble approach. (Color figure online)

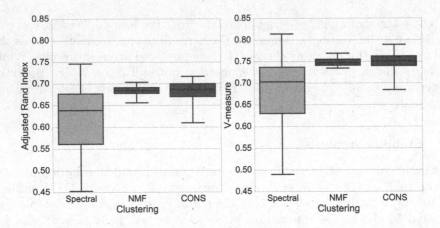

Fig. 5. Adjusted rand index and V-measure scores on 50 random subsampling experiments.

($p < 10^{-13}$ and $p < 10^{-12}$ for V-measure and adjusted rand index, respectively). Post-hoc Tukey test with 99% confidence reveals that both, NMF and CONS, significantly outperform results of individual clusterings, while difference in mean performances of the ensemble approaches are not significant.

4   Conclusion

The study presented here underscores the sensitivity of clustering results on chosen beta diversity measure that further leads to the uncertainties in the results interpretation. To avoid risk of choosing the less appropriate metrics and obtaining misleading or vague conclusions, we propose ensemble approaches in clustering. Ensemble clusterings produced stable results that highly surpassed average of the ensemble and were on the level of the best in the ensemble. These results were further confirmed by running experiments on random subsamples. NMF approach performed slightly better in terms of the lower variance compared to the CONS. Improved stability of the ensemble approaches comes at the price of larger computations. Ensemble clustering requires multiple runs of clustering algorithms under different input settings and that poses additional challenges for a large scale studies. Our future work will be extended to the environmental and soil metagenomics where data exceed TB size and good reference databases are missing. We need distributed computing solutions and highly parallel workflows for running such experiments.

Acknowledgements. This work was partly supported by Serbian Ministry of Education and Science (Project III 44006).

References

1. Zhou, J., He, Z., Yang, Y., Deng, Y., Tringe, Ş.G., Alvarez-Cohen, L.: High-throughput metagenomic technologies for complex microbial community analysis: open and closed formats. MBio **6**(1), e02288–14 (2015)
2. Mendoza, M.L.Z., Sicheritz-Pontn, T., Gilbert, M.T.P.: Environmental genes and genomes: understanding the differences and challenges in the approaches and software for their analyses. Briefings Bioinform. **6**(5), 745–758 (2015)
3. He, Y., et al.: Stability of operational taxonomic units: an important but neglected property for analyzing microbial diversity. Microbiome **3**(1), 20 (2015)
4. Kuczynski, J., et al.: Microbial community resemblance methods differ in their ability to detect biologically relevant patterns. Nature Methods **7**(10), 813–819 (2010)
5. Koren, O., et al.: A guide to enterotypes across the human body: meta-analysis of microbial community structures in human microbiome datasets. PLoS Comput. Biol. **9**(1), e1002863 (2013)
6. Yang, P., et al.: Microbial community pattern detection in human body habitats via ensemble clustering framework. BMC Syst. Biol. **8**(Suppl 4), S7 (2014)
7. Legendre, P., Cáceres, M.: Beta diversity as the variance of community data: dissimilarity coefficients and partitioning. Ecol. Lett. **16**(8), 951–963 (2013)

8. Ng, A., Jordan, M., Weiss, Y.: On spectral clustering: analysis and an algorithm. Adv. Neural Inf. Process. Syst. **2**, 849–856 (2002)
9. Brdar, S., Crnojević, V., Zupan, B.: Integrative clustering by nonnegative matrix factorization can reveal coherent functional groups from gene profile data. IEEE J. Biomed. Health Inf. **19**(2), 698–708 (2015)
10. Monti, S., Tamayo, P., Golub, T.: Consensus clustering: a resampling-based method for class discovery and visualization of gene expression microarray data. Mach. Learn. **52**(1), 91–118 (2003)
11. Caporaso, J.G., et al.: Moving pictures of the human microbiome. Genome Biol. **12**(5), R50 (2011)
12. Wilke, A., et al.: A RESTful API for accessing microbial community data for MG-RAST. PLoS Comput. Biol. **11**(1), e1004008 (2015)
13. Caporaso, J.G., et al.: QIIME allows analysis of high-throughput community sequencing data. Nature Methods **7**(5), 335–336 (2010)
14. Rideout, J.R., et al.: Subsampled open-reference clustering creates consistent, comprehensive OTU definitions and scales to billions of sequences. PeerJ **2**, e545 (2014)
15. Edgar, R.: Search and clustering orders of magnitude faster than BLAST. Bioinformatics **26**(19), 2460–2461 (2010)
16. DeSantis, T.Z., et al.: Greengenes, a chimera-checked 16S rRNA gene database and workbench compatible with ARB. Appl. Environ. Microbiol. **72**(7), 5069–5072 (2006)
17. Rosenberg, A., Hirschberg, J.: V-measure: a conditional entropy-based external cluster evaluation measure. In: EMNLP-CoNLL, vol. 7 (2007)
18. Hubert, L., Phipps, A.: Comparing partitions. J. Classif. **2**(1), 193–218 (1985)
19. Wagner, S., Wagner, D.: Comparing clusterings: an overview. Universität Karlsruhe, Fakultät für Informatik Karlsruh (2007)

Multilayer Data and Document Stratification for Comorbidity Analysis

Kevin Heffernan[✉], Pietro Liò, and Simone Teufel

Computer Laboratory, University of Cambridge,
15 JJ Thomson Avenue, Cambridge CB3 0FD, UK
{kevin.heffernan,pietro.lio,simone.teufel}@cl.cam.ac.uk

Abstract. In this work, we introduce two novel contributions to the study of comorbidity. The first is a new method for finding disease correlations, using a multitude of information sources. In the era of big data, methods such as evidence synthesis enable researchers to exploit many freely available information sources to enrich their analyses. This forms the basis for our method where in lieu of examining one form of evidence, we introduce a novel combination of sources, providing an indirect association between patient genetic data and the scientific literature. Our second contribution is a new method for stratifying the scientific literature when searching for newly discovered disease correlations. Given that the volume of published biomedical literature has increased dramatically, a clinician does not have the ability to read every relevant article. We therefore propose a new way for refining the literature search space to discover recently introduced disease correlations. Results show that our system can produce reasonable hypotheses for disease correlations, and that document stratification is an important aspect to take into account when using scientific literature.

Keywords: Multi-omics · Information retrieval · Text mining

1 Scientific Background

Comorbidity represents the presence of one or more diseases co-occurring with an index disease within the same patient [1]. Inherently tied with age and affected by factors such as clinical therapy, comorbidity is an important topic when addressing a patient's diagnosis, and can often have confounding effects on phenotypes such as the effect of treatments for breast cancer. Conventionally, disease correlation is analysed by employing information extraction techniques on a single type of data source, such as Electronic Health Records or claims reports [2,3]. However, in the era of big data the research community now has the opportunity to include many freely available information-rich sources into their analyses. Evidence synthesis is one such approach, where combining statistical evidence from multiple data sources can result in a stronger consensus, helping healthcare professionals make the most informed decision [4]. This forms the basis for our first proposed method, where in lieu of examining one source of evidence, we

© Springer International Publishing AG 2017
A. Bracciali et al. (Eds.): CIBB 2016, LNBI 10477, pp. 209–219, 2017.
DOI: 10.1007/978-3-319-67834-4_17

cultivate statistically significant data from multiple healthcare sources ranging from genetic information to scientific literature in order to more rigorously detect pairs of diseases which are highly correlated.

Integrating knowledge from scientific literature in addition to traditional bioinformatics resources such as a gene ontology is critical given that over the past decades, the volume of published biomedical literature has increased dramatically. PubMed (the US National Library of Medicine's literature service) provides access to more than 25 million citations, adding thousands of records daily. It has now become impossible for scientists to read all the literature relevant to their field. As a result, critical hypothesis-generating evidence is often discovered long after it was first published, leading to wasted research time and resources. For example, a number of papers which mention a strong disease correlation may have been published, but if they have a low citation count, the correlation might not yet be widely recognised by the community. This hinders the progress on solving fundamental problems such as understanding the mechanisms underlying diseases and developing the means for their effective treatment and prevention. To address this problem, an entire domain of research called Literature Based Discovery has emerged as a prominent field of study, where links between biomedical articles are established for potentially undiscovered solutions to well known diseases.

When searching the scientific literature for new disease correlations, only citation counts have primarily been used in previous comorbidity studies. The intuition for concentrating on citations is that given the large volume of scientific literature constantly being released, a low citation count is a good indicator for poor information diffusion. Poorly cited papers can contribute important new hypotheses which haven't been discovered by the wider community. Often a low citation count can also be contributed to a lower quality publication, but recency of publication also plays a large role (i.e. new papers will need time to garner a high citation count). However, papers from well known journals gather high publication counts very quickly, and may also contain important new hypotheses. Our method therefore allows to use all hypotheses from high and low cited papers.

We also advocate to directly use the context of where hypotheses were found. For example, a review paper is not likely to introduce many new concepts into the community, but an experimental paper may show novel, previously undiscovered links between entities. We consider two factors, namely the journal category and publication type of an article. We make a distinction on how much importance is to be placed upon each journal category or publication type from the information sources being collated. This forms our second proposed method, which we call *document stratification*.

2   Materials and Methods

2.1   Multilayer Data

The main information source for our first method is multilayer data. This section describes each layer and how the inter-layer correlations were calculated.

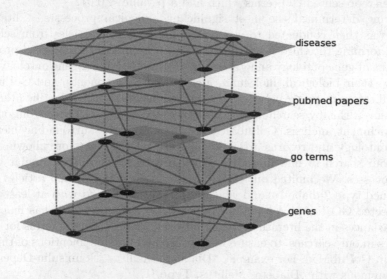

Fig. 1. Schematic diagram of our multilayer data. The creation of the multilayer data begins with a base layer of genetic data, and results in the top layer containing our disease correlation hypotheses. Each layer's nodes are connected to nodes in adjacent layers by weighted ties.

The construction of our multilayer data begins with a base layer of gene expression data from patients with an index disease of interest, providing over and under expressed genes with which to pursue enrichment. We chose type 2 diabetes as our index disease, given that this is a well studied condition in the comorbidity literature. To extract the gene data, we chose to use GEO accession GSE38642[1]. This data set contains human pancreatic islets from 54 non-diabetic and 9 diabetic donors. Human pancreatic islets are especially appealing for an analysis concerning type 2 diabetes as these islets contain beta cells which secrete insulin, giving them a significant role in this condition. Care was taken not to confound our results due to patient differences such as age or gender, and so we chose a subset of the dataset population. This subset contained females who are under 50, therefore allowing us to study a stratified population who also have a reduced chance of comorbidity due to aging. To ensure these patients were comparable, we calculated the value distribution. This distribution was approximately median-centered across samples, and therefore valid for comparison. For

---

[1] https://www.ncbi.nlm.nih.gov/geo/query/acc.cgi?acc=GSE38642.

gene expression analysis, the R package Limma [5] was chosen, and from the expression output we selected all over- and under-expressed genes which had a p-value < 0.01.

These up- and down-regulated genes were then used as input to construct the second layer of data, where following gene collection, we performed a gene set enrichment analysis. Following the gene set enrichment, we extracted under and over-represented GO terms which had a p-value < 0.01.

Having determined the most significant biological processes, a literature search was then conducted to assemble a collection of articles from scientific papers, forming the third layer. These articles are linked to biological processes by means of gene ontology annotations in GoPubMed [6]. A gene ontology maps genes to their biological, molecular, and cellular components, and GoPubMed can facilitate a mapping between these components and the scientific literature. We achieved this by searching articles annotated with GO terms from our earlier enrichment analysis. GoPubMed also examines the surrounding hierarchy in the ontology and returns articles which are related, therefore allowing us to widen our search to find literature which involves functionally similar biological processes. We limited our search space to that of Pubmed articles which mentioned type 2 diabetes, and which were also annotated with at least one of our selected GO terms. Additionally, given that a disease name has many possible variances in the literature, we included all MeSH [7] entry terms for type 2 diabetes in our searches, to ensure we captured all possible mentions of this condition in GoPubMed. For example, "Diabetes Mellitus, Noninsulin-Dependent" is synonymous with "Diabetes Mellitus, Type II".

The final data layer contains our disease hypotheses, gathered using information extraction on the articles collected in the literature layer. Diseases which were present in the collected literature were extracted using MetaMap [8]. MetaMap is a tool which is able to map text in the biomedical domain to concepts stored in the UMLS Metathesaurus [9]. For each Pubmed article, we stored which diseases were present, and also how many times each disease was mentioned.

This data collection resulted in four layers containing nodes of size {217, 17, 899, 683} respective to gene, GO term, Pubmed and disease data collection. A schematic overview of this data collection process can be seen in Fig. 1.

2.2   Document Stratification

To tackle the problem of a large search space in information retrieval arising from ever growing publications, we propose a two-tier weighted stratification of the scientific literature. The first level of stratification concerns the category of journal which the article was sourced from. When searching for comorbidity, there are many journals which are likely to mention disease co-occurrences, but which should not be emphasised due to the nature of their material. For example, consider the *International Journal of Legal Medicine*. This journal falls under the category of *Medicine (Legal)*, and is not a journal which a medical practitioner would make use of when researching comorbidity. Therefore, journals such

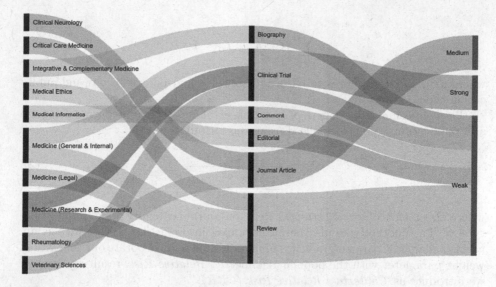

Fig. 2. Example of 2-tier document stratification showing journal category (left), publication type (centre), and resulting correlation strength (right).

as these clearly adds noise given that many journals contain topics completely unrelated to practical medicine.

The second tier of stratification makes use of the varying publication types which an article may be categorised as. In addition to the journal category, different types of publications are likely to contain more novel discoveries than others. For example, a Clinical Trial may show new links between diseases, while an Editorial or Biography are articles not likely to contain pertinent information regarding knowledge discovery.

An example of this two-tier system can be seen in Fig. 2. The correlation strengths are determined by the paths which were taken via journal category, and publication type respectively. Consider the following example: an article is found, and the publication type is Clinical Trial. This type of publication is weighted strongly for new disease correlations. However, this article came from a journal in Veterinary Sciences. Therefore, even though the material is likely to show new links between entities, the subject matter is not applicable, and so the resulting correlation strength determined from this article is *weak*. However, an article from a journal in the Medicine (Research & Experimental) category with a publication type of Clinical Trial results in a *strong* correlation strength, since both of these paths are weighted favourably for new disease correlations.

The weights for journal category and publication type can be freely chosen by the clinician. This will result in a series of independent weights for each stratification tier with their product defined as:

$$\sum_{\alpha=1}^{n} JC_\alpha \sum_{\beta=1}^{m} PT_\beta = 1 \tag{1}$$

where JC_α and PT_β represent chosen weights for journal category α and publication type β respectively.

The contextual risk strength of an article given its associated journal category and publication type is formalised using the formula:

$$\Omega_{\alpha\beta} = \frac{JC_\alpha PT_\beta}{\sum_{i=1}^{n} JC_i \sum_{j=1}^{m} PT_j} \tag{2}$$

where $\Omega_{\alpha\beta}$ is the contextual strength between journal category α and publication type β. This method can also be incorporated into existing risk metrics as an additional weight parameter. An example would be to incorporate our contextual weight parameter with the popular risk metric, *Relative Risk*, resulting in what we introduce as *Contextual Relative Risk (CRR)*:

$$CRR_{ij,\alpha\beta} = \frac{C_{ij}/N}{P_i P_j/N^2} \cdot \Omega_{\alpha\beta} \tag{3}$$

where C_{ij} are the number of articles which contain diseases i and j, N is the total number of articles, and P_i and P_j are the probabilities of finding diseases i and j respectively.

To see how disease correlations were being affected by document stratification, we collected two samples of articles from Pubmed grouped by publication type and journal category respectively. We chose to examine two publication types (Clinical Trial, Review), and four journal categories from the field of medicine (Research & Experimental, Legal, Informatics, Ethics). A disease vector was created for each article which contained all disease mentions identified by MetaMap. For each disease vector, we computed all possible unique disease pair combinations.

3   Results

3.1   Multilayer Data

In order to see if our novel method for predicting comorbid diseases was producing hypotheses which were reasonable, we compared our results against the largest publicly available dataset of disease correlations, HuDiNe [10], which contains hospital claims data for over 13 million patients.

For evaluation of our hypothesis generation system against HuDiNe, we chose to compare the top ranked conditions from both systems. Two of the metrics used by HuDiNe to rank disease correlations are the phi correlation and t-value. To calculate both metrics, we searched the entirety of Pubmed articles for disease correlation. Pubmed contains citations from over 25 million articles, therefore

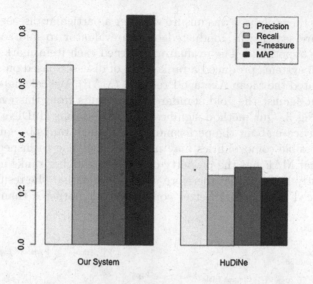

Fig. 3. Results from expert annotator evaluation of complex network data. Our system significantly beats HuDiNe across all measures.

making it an excellent corpus with which to rank our hypotheses. The phi correlation between diseases i and j is defined as:

$$\varphi_{ij} = \frac{C_{ij}N - P_iP_j}{\sqrt{P_iP_j(N - P_i)(N - P_j)}}$$

where C_{ij} is the number of articles which contain both disease i and j, N is the total number of articles available in Pubmed, and P_i and P_j contain the number of articles which mention diseases i and j respectively. To determine the significance of each comorbid hypothesis, we then estimated a t-value for each phi correlation, which is defined as:

$$t_{ij} = \frac{\varphi_{ij}\sqrt{n - 2}}{\sqrt{1 - \varphi_{ij}^2}}$$

where n is $\max(P_i, P_j)$.

To evaluate our hypothesis generation system against HuDiNe, we compared the top ranked conditions from both systems, using an expert evaluator from the European Bioinformatics Institute. In order to prevent bias in the evaluator's decision, care was taken so that the presentation of the diseases given for annotation would not identify which system they were from. This was done by first randomising the ranking of each list separately, and then creating an amalgamated list of both system's outputs by selecting a disease from each list in an interleaving fashion. For each hypothesis for a comorbid disease, the evaluator was asked to determine whether each output hypothesis was reasonable.

In cases where the evaluator was unsure whether a particular disease was comorbid, a literature search was conducted by the evaluator to find sources which supported the hypothesis. The evaluator returned each item marked as Yes or No. Since both systems produced a ranked list of diseases based on significance, we also calculated the Mean Average Precision (MAP). We calculated the MAP of each system against the gold standard. The results from this evaluation are presented in Fig. 3. Our method significantly outperforms HuDiNe ($p < 0.05$). We particularly care about the performance in the high ranks because clinicians looking for possible comorbidities may not have the time to inspect long lists. Please note that MAP has the property of weighting higher ranks more heavily. Our significantly higher MAP therefore also indicates that the results from our system produced significantly better comorbidity hypotheses than HuDiNe in the top ranks.

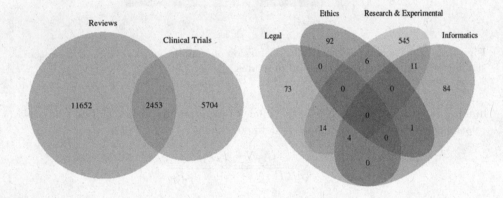

Stratification type 1 (Publication Type)
Disease pairs from clinical trial and review articles.

Stratification type 2 (Journal Category)
Disease pairs from articles in four journal categories from medicine: Research & Experimental, Legal, Informatics and Ethics.

Fig. 4. Disease pair occurrence and overlap for subset of Pubmed articles stratified by examples of publication type (left) and journal category (right).

3.2 Document Stratification

As can be seen in Fig. 4, there is some overlap of disease pairs for publication type, and journal category e.g., *Reviews* ∩ *Clinical Trials* = 2453. There are also a considerable number of disease pairs which are mentioned in one category only (non-overlap). There are many disease pairs which are only referenced in Clinical Trial articles, not in Reviews, therefore showing possibly novel disease pairings. Regarding publication type, the existence of non-overlap makes an even stronger case for upweighting the combination coming from Clinical Trial papers.

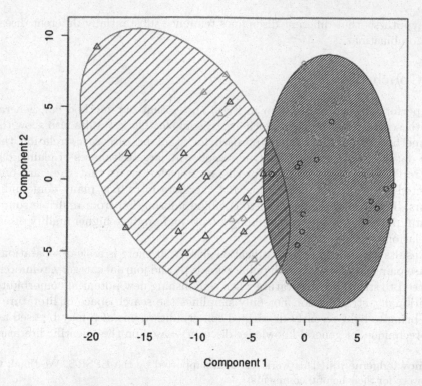

Fig. 5. k-means clustering of disease pairs from clinical trial and review articles using cosine distance of disease pair names. Black and green nodes denote clinical trial and review publication type membership respectively. Various values for k were chosen, with k = 2 giving the best silhouette coefficient. (Color figure online)

The journal category emphasises the need for placing weaker weights on categories such as Medical Ethics papers. The majority of disease pairs mentioned here are not referenced in a medical research journal, and so taking these into account will only add uncertainty, i.e. "noise" to comorbidity analysis.

In the case of publication type, we took a look at how closely related the disease pairs were. If it were the case that different publication types differed in the disease pairs they contained, we would expect some overlaps and some non-overlaps to emerge from clustering, as in the left-hand side of Fig. 4. To calculate disease pair similarity, we created a string distance matrix for all disease pairs. This matrix was calculated using cosine distance, which we then clustered using k-means (k = 2, purity = 0.78). The number of clusters was chosen by clustering using various values for k, and choosing the k with the highest average silhouette coefficient (0.51). This clustering is shown in Fig. 5. The left cluster shows a clear separation between disease pairs in Clinical Trials, and those described in Reviews. This indicates that the non-overlapping disease pairs in Clinical Trial articles are substantially different to those referenced in Reviews, and confirms our hypothesis that different disciplines reference substantially different disease pair combinations.

4   Conclusion

The performance of our system in Fig. 3 shows that our method can generate hypotheses for comorbidities which are reasonable. These results also show that our method may be a better approach for studying disease correlation than large databases of claims data. One reason why large databases of claims data are troublesome when looking for disease correlations is that they are often collected from an aged population and therefore introduce many confounding factors. This reinforces the need to combine information from multiple sources of data in order to reduce confounding factors and produce higher quality disease correlations.

Results from document stratification show that there is a clear separation of disease pairs mentioned in each publication type and journal category, reinforcing the need to stratify the literature when establishing new potential comorbidites. Additionally, stratification not only simplifies the search space of literature to help bring to light recently discovered comorbidities, but also may be used as a new technique for general knowledge discovery based on the scientific literature.

Acknowledgements. This work has been supported by the EPSRC. We thank the reviewers for their helpful comments.

References

1. Valderas, J.M., Starfield, B., Sibbald, B., Salisbury, C., Roland, M.: Defining comorbidity: implications for understanding health and health services. Ann. Family Med. **7**(4), 357–363 (2009)
2. Ware, H., Mullett, C.J., Jagannathan, V.: Natural language processing framework to assess clinical conditions. J. Am. Med. Inform. Assoc. **16**(4), 585–589 (2009)
3. Salmasian, H., Freedberg, D.E., Friedman, C.: Deriving comorbidities from medical records using natural language processing. J. Am. Med. Inform. Assoc. **20**, e239 (2013). amiajnl-2013
4. Sutton, A.J., Welton, N.J., Cooper, N., Abrams, K.R., Ades, A.E.: Evidence Synthesis for Decision Making in Healthcare, vol. 132. Wiley, Hoboken (2012)
5. Smyth, G.K.: Limma: linear models for microarray data. In: Gentleman, R., Carey, V.J., Huber, W., Irizarry, R.A., Dudoit, S. (eds.) Bioinformatics and Computational Biology Solutions Using R and Bioconductor. Statistics for Biology and Health, pp. 397–420. Springer, New York (2005). doi:10.1007/0-387-29362-0_23
6. Doms, A., Schroeder, M.: Gopubmed: exploring pubmed with the gene ontology. Nucleic Acids Res. **33**(suppl 2), W783–W786 (2005)
7. Lipscomb, C.E.: Medical subject headings (mesh). Bull. Med. Libr. Assoc. **88**(3), 265 (2000)
8. Aronson, A.R.: Effective mapping of biomedical text to the UMLS metathesaurus: the MetaMap program. In: Proceedings of the AMIA Symposium, p. 17. American Medical Informatics Association (2001)

9. Bodenreider, O.: The unified medical language system (UMLS): integrating bio-medical terminology. Nucleic Acids Res. **32**(suppl 1), D267–D270 (2004)
10. Hidalgo, C.A., Blumm, N., Barabási, A.L., Christakis, N.A.: A dynamic network approach for the study of human phenotypes. PLoS Comput. Biol. **5**(4), e1000353 (2009)

Evolving Dendritic Morphologies Highlight the Impact of Structured Synaptic Inputs on Neuronal Performance

Mohammad Ziyad Kagdi[✉]

Mental Health, Clinical Research Centre, University of Aberdeen,
Cornhill Road, Aberdeen, Aberdeenshire AB25 2ZH, UK
m.ziyad@kagdi.org

Abstract. Dendrites, the most conspicuous elements of neurons, extensively determine a cell's capacity to recognise synaptic inputs. Investigating its structure and morphological properties unravels the functioning mechanism of neurons that cooperates the process of learning and memory. This research systematically generates a varying topology of dendrites in a multi-compartmental model of a neuron with passive properties and it further explores a cell's integration ability of complex synaptic potentials. The neurons receive an equal number of binary input patterns of synaptic activity and the performance of a cell is gauged by calculating the signal to noise ratio between amplitudes of somatic voltage. The objective is to analyse the types of input pattern in combination with morphological properties that may strengthen or weaken the somatic response. Finally, an evolutionary algorithm produces a fine variety of branching structures calculating the weighted sum of synaptic inputs, further identifying the impact of membrane and morphological properties on neuronal performance.

Keywords: Dendritic morphology · Synaptic integration · Synaptic plasticity · Hebbian learning · Pattern recognition · Evolutionary algorithm

1 Introduction

A neuron is a nerve cell excited electrically to process and transmit information through electrochemical signals. Many different types of neurons exist in the human brain - approximately 10^{11} to 10^{12} in number - with a great variety of morphologies. Each neuron connects on average to 10^{14} other neurons, constituting a total of 10^{15} to 10^{16} connections known as synapses [14], thereby establishing the broader realm of human perception, emotions, thoughts and memories. There are four typical components of a single neuron; dendrites, soma, axon and axon terminals. Dendrites are tree like branching structures, often extended away from soma (cell body) for hundreds of micrometers (μm), play an important role in propagating and integrating synaptic potentials [2,13]. Soma is the

© Springer International Publishing AG 2017
A. Bracciali et al. (Eds.): CIBB 2016, LNBI 10477, pp. 220–234, 2017.
DOI: 10.1007/978-3-319-67834-4_18

neuronal cell body attached to dendrites, containing nucleus and other cellular components, and is responsible for sending and receiving electrochemical signals.

The development of dendrites is regulated by intrinsic genetic factors and cellular organisation of actin and microtubule cytoskeleton for the formation of pertinent dendrite morphology [10]. Dendrites considerably vary in their anatomical structure and are thought to have associated with variety of computational tasks. Dendritic spine is a microscopic membranous protrusion on neuron's dendrite, containing postsynaptic compartment of excitatory synapse to serve as a storage site for synaptic strength [10]. Transmission of electrical stimulations from other neurons is carried out via microscopic junctions called synapses which are located at the various points (spines) across the dendritic arbor. The primary interest of studying the human brain lies in exploring the information processing mechanism among different types of neuronal morphologies. Neurons show a lot of variability in their shape and structure, exhibit a wide variety of patterns and strength of connections through which memories are stored and habits are learned. There have been many investigations done on neurons with different types and morphologies, yet the reason behind these varieties and functional implications of different morphologies remains unclear. Few studies have hypothesised that the variability in morphological structures is unlikely to be accidental and that these variabilities could exist due to optimised propagation of neuronal signals from synapses to soma [2,13]. Synaptic integration is a complex process that comprises a great deal of computations within dendrites, and requires concurrent inputs from excitatory synapses to determine neuronal firing behaviour. It is one of the possible roles of neuronal dendritic arborization to recognise structured synaptic inputs through integrating various arriving signals at the soma.

The present study focuses on the development of dendritic arbors using partition notations to mimic its tree-like structure and it further evaluates the impact of different structures on a cell's pattern recognition capacity. The developmental approach used here for dendritic growth is based on two different methods. Firstly, a branching stochastic approach is used with partition notations [12] to generate a possible number of dendritic branches and then, an evolutionary algorithm (EA) is utilised to produce an optimum dendritic structure suitable for recognising synaptic inputs. A neuron receives input signals from many other neurons attached to it and the strength of those signals is defined by its synaptic plasticity. To characterise a neuron based on its passive properties, a compartmental model [7] is utilised to simulate the postsynaptic integration of excitatory inputs. The compartmental model of a neuron divides each complicated dendrite into number of compartments and imitates the behaviour of a biologically realistic neuron. On the contrary, an artificial neural network model (ANN) of an associative memory, implementing the Hebbian learning rule is also employed to calculate the weighted sum of its synaptic inputs, thereby comparing its pattern recognition performance with that of the compartmental model neuron. To optimise a neuron's information processing capability, some binary patterns called stored patterns, representing synaptic inputs are presented during a learning phase. Once the learning phase is completed, some novel binary patterns are presented to discriminate them from the stored patterns. A neuron's discriminatory ability is dependent upon its dendritic structure, spatially distributed excitatory inputs and strength of connections as measured by the signal to noise ratio (S/N) between the somatic EPSP amplitudes/weighted sum responses of the stored and novel patterns. The higher the S/N ratio, the better able the cell to discriminate the two sets of input patterns. More importantly, it is also hypothesised that the spatial distributions of synaptic inputs could affect the postsynaptic response, as a result, it may strengthen or weaken the neuronal performance by sending signals to the soma from shorter or longer dendritic distance. To identify whether this is correct, current study also focuses on presenting some biased synaptic input patterns called fixed stored patterns in which input locations on dendrite are manually determined. Additional parameters such as axial resistivity, compartment and mean path length, temporal asynchrony of signal arrival are also investigated to ascertain whether these parameters have any association with the cell's pattern recognition performance.

2   The Model

The primary aim of this study is to understand the functioning mechanism of a neuron that cooperates the process of learning and memory by storing and recognising synaptic inputs. To begin with, neuron models are presented with membrane and morphological properties (Sects. 2.1 and 2.2), after which an EvOL-DnDR[1] algorithm generates a population of 100 neuron models with diversified dendritic topologies, which are further provided with the sets of synaptic inputs and their performances are measured by calculating the S/N ratio.

2.1   Passive Membrane Properties

A passive neuron model is used to understand its electrical properties without any active conductances in the soma and dendrites, therefore it does not generate action potentials. From the study carried out by De Sousa [3], the following passive parameters are considered for membrane capacitance, membrane resistance and axial resistivity, respectively: $C_m = 0.75 \ \mu F/cm^2$, $R_m = 30 \ k\Omega cm^2$ and $R_a = 150 \ \Omega cm$.

2.2   The Compartmental Model of a Neuron

In the compartmental model, a neuron is treated as a cell body with divided isopotential dendritic compartments, receiving input signals in the middle of every compartment as shown in Fig. 1(C). The charge across each compartment is same and can be represented by an electrical circuit. The length and diameter of soma is based on the study carried out by De Sousa [3] where the soma

---

[1] http://research.kagdi.org/cns/evolving-dendritic-morphologies.

is a cylinder of 20 μm length and diameter, followed by 500 μm length and 25 μm diameter for each dendritic compartment. The conductance amplitude of a naive synapse is set to 1.5 ηS (before learning) and is subject to change once multiplied with the weight value after synaptic learning. The resting membrane potential of a model neuron was set to −65 mV before the transmission of any excitatory inputs. On arrival of these input signals, an excitatory postsynaptic potential (EPSP) amplitude, representing the depolarisation of membrane volt-age is generated. An EPSP is the somatic excitation needed for any neuron to fire an action potential (AP), caused by the incoming active signals.

2.3   Generation of Dendritic Structures

To define the growth of a multi-compartmental dendrite, partition notations are used involving axioms and rules to constitute the branching structure [12]. In a binary tree, a partition at the bifurcation point is defined by a pair of numbers denoting the degree of each subtree, where the terminal nodes in each subtree are further divided into left and right branches as tree grows. Therefore, a sequential specification containing partition notations executes in a linear order to bifurcate and generate branches of a tree. For instance, a tree with 4 terminal nodes and 7 compartments can be specified using the partition notations 4(2(1,1)2(1,1)) and 4(1 3(1 2(1,1)) as available in Fig. 1(A) & (B).

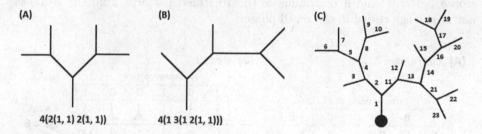

Fig. 1. Partition notations used to define dendritic trees with 3 terminal nodes and 7 compartments in (A) & (B), and Branching of a neuron using the compartmental model in (C).

In the present study, a population of 100 neurons is generated with T = 128 terminal nodes and 255 (= 2T − 1) compartments, introducing different types of dendritic topologies by stochastically determining number of terminal points at every branch in a tree.

2.4   Synaptic Plasticity and Long-Term Potentiation

A learning process in neurons demonstrates the cells' activity-dependent adaptive behaviour characterised by the Hebbian learning rule of growing synaptic strength between the firing neuron and the neuron receiving input signals called synaptic plasticity [6]. A change in synaptic strength occurs due to persistent stimulations by pairs of pre and post synaptic neurons, where a persistent increase in synaptic strength represents Long-term potentiation (LTP) [11]. LTP is an input-specific process shaped by pre-synaptic activations which as a result determines the behaviour of a postsynaptic neuron. To achieve the synaptic strength in simulated neurons, a weight value is used that increases by 1, every time there is an active synapse established between the pre and post synaptic neurons, which consequently stores a history of active connections representing neuronal activities, as shown in Fig. 2(A).

2.5 The Pattern Recognition Task

The ability of a neuron to discriminate between patterns of synaptic inputs is largely dependent upon its ability to recognise number of active synapses [4]. Comparison is made between the performance of a computing unit in an artificial neural network (ANN) model of an associative memory and the compartmental model of a biologically realistic neuron. At the outset, two sets of 10 binary synaptic input patterns, called the stored and the novel patterns are provided to the computing unit in ANN. Each pattern contains 255 bit series of randomly generated binary values representing active and inactive synapses (where 1 denotes an active synapse) presented to each unit's 255 input layers. There are 25 active synapses in each pattern. Under the training phase, the unit learns 10 stored patterns, and it discriminates the 10 stored patterns from the 10 novel patterns when tested in the recall phase.

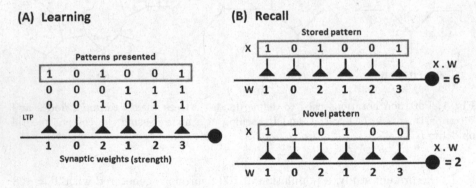

Fig. 2. LTP learning with 3 synaptic input patterns in (A), and a Recall phase with synaptic weights to recognise patterns in (B).

On each active synapse from the stored patterns, a typical weight value is increased by 1, describing the concept of synaptic strength, as shown in Fig. 2(A). A response to a specific pattern is a sum of all inputs multiplied by the associated synaptic weights (Dot product of an input and weight value) representing the dendritic summation, as calculated for both the pattern sets using Eq. 1 and is shown in Fig. 2(B). The dendritic sum of an output unit is considered as an ultimate response to each pattern presented, and it is used to calculate the S/N ratio using Eq. 2 [3].

In the secondary aspect of the pattern recognition task, both sets of input patterns along with the associated synaptic weights are transferred to the NEURON compartmental model [7]. In a population of 100 compartmental model neurons with varying dendritic morphologies, the 255 binary inputs are spatially distributed to each neuron's 255 dendritic compartments (synapses), each active input arrives synchronously at its associated synaptic location. The resultant EPSP amplitudes of each compartmental model neuron for both the pattern sets were recorded. The EPSP amplitudes represent a change in each cell's somatic voltage, showing the higher and lower membrane potentials, are considered as a response to the sets of input patterns provided. There is a strong association between the somatic EPSP amplitudes and a cell's possible AP [4], indicating that the EPSP(s) play a crucial role in determining the cell's pattern recognition performance.

(A) EPSP Responses   **(B) Frequency of the Peak Responses**

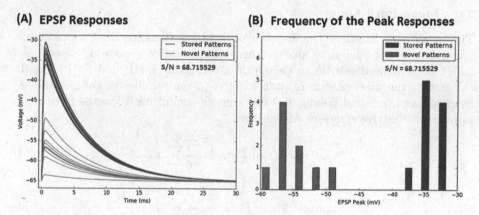

Fig. 3. The EPSP amplitudes of passive neuron to 10 stored and 10 novel patterns in (A), and the frequency of the peak EPSP responses to both stored and novel patterns (bin width = 1 mV) in (B). (Color figure online)

The generated EPSP amplitudes were used to calculate the S/N ratio between the responses of stored and novel patterns, which later compared with the performance of a computing unit in an ANN model. Various forms of noise that degrade spatio-temporal integration of synaptic inputs are absent in an ANN model, which are investigated in the compartmental model neurons. A high performing neuron was thus identified as the one with maximum S/N ratio in a population of 100 compartmental model neurons, discriminating between the sets of synaptic inputs, as shown in Fig. 3(A). The following equations are used to calculate the Dendritic Sum and the S/N ratio.

$$Dendritic\ Sum = \sum_{i=1}^{n} X_i.W_i \ . \tag{1}$$

In Eq. 1, n refers to the number of synaptic compartments/weights which is 255 for each binary pattern, X_i is the ith input signal, W_i is the ith weight value in the associated synaptic pattern, as shown in Fig. 2(A) and (B).

$$S/N = \frac{(\mu_s - \mu_n)^2}{0.5(\sigma_s^2 + \sigma_n^2)} . \tag{2}$$

In Eq. 2, μ_s and μ_n are the mean values and σ_s^2 and σ_n^2 are the variances of the ultimate/peak responses of stored and novel patterns. Graham proposed a similar type of LTP learning model, calculating the S/N ratio and discriminating the synaptic inputs in the presence and absence of noise in a multi-compartmental model of a CA1 hippocampal pyramidal neuron. A study which concluded that the amplitudes of voltage responses at the soma is dependent upon the spatial distribution of synapses and that the variations in amplitudes occur due to different synaptic locations (which as a result cause the temporal asynchrony of signal arrival at the soma) even with the same number of active synapses [4].

2.6   Mean Path Length

The mean path length is the measure of dendritic distance from soma to terminal points, considered to analyse the performance of a neuron. Equations 3 and 4 denote the dendritic and synaptic mean path (DMP and SMP) length, calculating the average sum of path lengths (mean number of compartments) from soma to terminal points, and the average sum of path lengths from soma to number of active synapses respectively.

$$Dendritic\, Mean\, Path = \frac{1}{n}\sum_{k=1}^{n} P_k . \tag{3}$$

$$Synaptic\, Mean\, Path = \frac{1}{n}\sum_{k=1}^{n} Q_k . \tag{4}$$

Here, P_k is the length of dendritic path to the kth terminal point, and Q_k is the length of dendritic path to the kth synaptic compartment from the soma. Figure 4 shows the calculated DMP and SMP length for each tree available, where a red line represents an active synapse to that particular compartment.

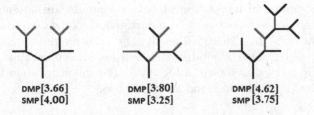

DMP[3.66]   DMP[3.80]   DMP[4.62]
SMP[4.00]   SMP[3.25]   SMP[3.75]

Fig. 4. Dendritic and synaptic mean path (*DMP and SMP*) lengths are calculated.

2.7 Evolutionary Algorithm and Exploration of Dendritic Structures

An evolutionary algorithm (EA) is a heuristic optimisation technique inspired by the Darwinian theory of survival of the fittest and natural genetics, applied to a population of individuals for breeding high quality solutions [9]. In the present study, an EA is implemented with a population of 100 candidate solutions (compartmental neurons) to produce a fine variety of branching structures and to maximise the S/N ratio. It is the secondary approach of producing some functionally desirable dendrites in each generation by only selecting high performing neurons (with maximum S/N ratio) for reproduction. Once pairs of individuals are selected, the genetic features of dendrites are exchanged using dual point crossover and offspring are mutated with 4% to 20% of a mutation rate. Additionally, the concept of elitism is also utilised, keeping about 10% to 15% of best individuals intact in the next generation without any genetic modifications. Elitism makes sure that the EA does not lose high performing neurons once their genetic details are exchanged or mutated to produce future offspring. The above steps are repeated until some most fit neurons are reproduced.

3 Results

3.1 Distortion of Input Signals in Dendritic Trees

Membrane potential plays a crucial role in determining neuronal performance, and therefore it is essential to identify the causal relationship between the travelling inputs and the attenuation of an EPSP response. On a dendritic tree, individual synapses may differ in their distance from soma. As a result, the amplitude of voltage response at the soma can be affected by the spatial distribution of these input signals as they travel along the lengthy dendritic tree to the final integration site. Figure 5 shows examples of voltage responses generated at the soma due to different distribution of synaptic inputs (blue traces) on the dendritic tree, whereas variations in voltage amplitude occur, such as an EPSP amplitude decreases and its time span lengthens, with the distance of an originating synapse from the soma. An EPSP response of 22 sparsely attached active inputs to the 22 distal dendritic compartments in Fig. 5(A) & (B) shows only about 0.5 mV of postsynaptic depolarisation as opposed to 8.5 mV when inputs are clustered proximally to the soma, see Fig. 5(C) & (D). Interestingly enough, despite that the distributed excitatory inputs arrive synchronously at individual synapses, the spatio-temporal integration of transient signals is still disrupted due to asynchronous arrival of excitatory potentials at the soma, significantly affected by the varying synaptic distances from the cell body. Figure 5(B) & (D) shows calculated temporal asynchrony of signal integration and the synaptic mean path length (SMP) depicting larger value for distorted and delayed membrane voltage. Similarly, the higher R_a of individual compartment increases the intracellular resistance for ions to move and disrupts the flow of synaptic potentials. Hence, the amount of signal attenuation witnessed here is proportional to

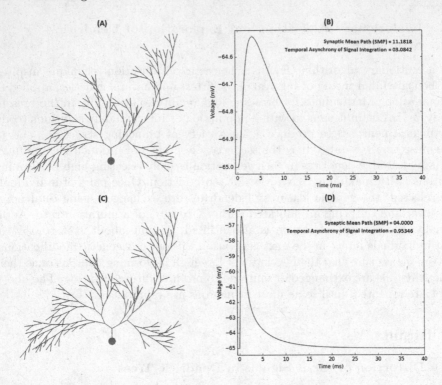

Fig. 5. 22 active synaptic inputs are attached to compartments at distal/proximal to the soma in (A) & (C) and their voltage responses in (B) & (D) with calculated Synaptic Mean Path (SMP) and Temporal Asynchrony of Signal Integration. (Color figure online)

the value of axial resistivity used ($R_a = 120$ Ωcm), showing inverse association with degrading membrane potential.

3.2 Discrimination of Synaptic Inputs and the Signal to Noise Ratio

Before measuring the somatic EPSP responses of a compartmental model neuron, two sets of randomly generated synaptic input patterns were provided to the computing unit in an ANN model and their dendtric sums were calculated using Eq. 1; see frequency distribution in Fig. 6(A). The s/n ratio was 40.1993 generated from the dendritic sums of both the pattern sets. Although, the distinction is quite clear, it is important to measure the somatic response in the compartmental model neuron. Once measured, the resultant S/N ratio was 10.130346 with some rather overlapping EPSP amplitudes (results not shown). The DMP and SMP lengths were 9.531 and 8.448, calculated using Eqs. 3 and 4 respectively. Due to the wider distribution of input synapses, the S/N ratio was relatively lower, showing poor discrimination of stored and novel patterns. To produce some

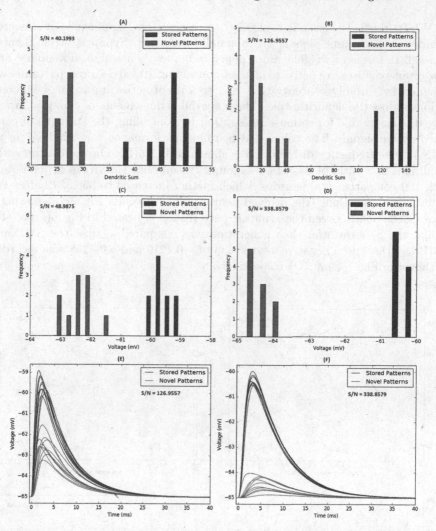

Fig. 6. Evolution of dendrites for recognising random and clustered (fixed) synaptic inputs, showing frequency distribution of an ANN model's weighted sums in (A) & (B), frequency distribution of an evolved neuron's peak responses in (C) & (D), and their associated EPSP amplitudes in (E) & (F) respectively.

high performing neurons, an EA was utilised, evolving 100 individuals for 100 number of generations, which finally produced an optimised neuron (Appendix Fig. 9A) showing discriminatory EPSP amplitudes with its improved S/N ratio of 48.987556, as shown in Fig. 6(C) & (E). Performance of a most fit neuron in each generation was compared against its SMP and DMP length which showed a negative correlation of −0.7937 and −0.4947 respectively, indicating growing improvements in the neuronal performance (S/N ratio) with closer synapses to the soma, see Fig. 7(A) & (B).

Whilst this is the case, it is also key to measure the significance of clustered synapses, when input signals are manually arranged. Synaptic strengthening of a cell in forms of weight values depends however on where, how many and how often synapses are activated. A set containing 10 fixed stored patterns was designed by variably concentrating 20% to 25% of active inputs at 4 different regions across the dendritic tree. These spatial arrangements of synaptic inputs played a major role for somatic integration by controlling the strength of combined active signals. The calculated neuronal performance was 126.9557 in an ANN model, frequency distribution is shown in Fig. 6(B). On the contrary, the maximum S/N ratio of 83.5114 was found (results not shown) in a population with 100 compartmental neurons, which further increased to 338.8579 after 100 evolutionary iterations (the evolved neuron is in Appendix Fig. 9B), showing a clear distinction between the synaptic patterns, as indicated in Fig. 6(D) & (F). Again, the S/N ratio for each generation was compared against the SMP and DMP lengths, and a negative association of −0.4210 and −0.3258 was recorded as shown in Fig. 7(C) & (D) respectively.

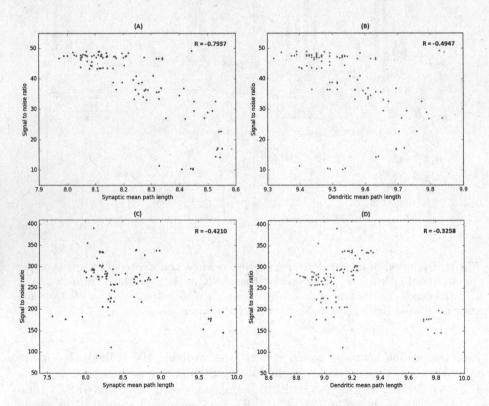

Fig. 7. The S/N ratio of a most fit neuron from each evolutionary generation was compared against its SMP and DMP length, recognising both the random and clustered input patterns in (A), (B) & (C), (D) respectively.

3.3   Temporal Asynchrony of Somatic Integration

Since the random and a wide distribution of synaptic inputs show a great variety of distances from soma, there exists a temporal asynchrony of signal arrival at the final integration site. An effect of which broadly affects the neuronal response by making it highly inhomogeneous, as shown in Fig. 5. Therefore, it is also likely that the variety of these EPSP responses are proportional to the temporal irregularity of incoming inputs. It would be useful to ascertain whether these temporal irregularities of signal arrival have a causal relationship with their associated inhomogeneous somatic responses. An experiment was carried out in which an EA recorded the temporal irregularity of signal arrival for each most fit neuron for 100 generations. Once they were compared with the varying peak responses of stored patterns, a positive association of 0.631279039 was found, suggesting growing variations of somatic responses with increasing temporal asynchrony of signal arrival at soma. Similarly, the temporal asynchrony was also compared against the measure of s/n ratio which depicted a negative correlation of −0.5151, showing poor neuronal performance with a growing synaptic irregularity, see Fig. 8.

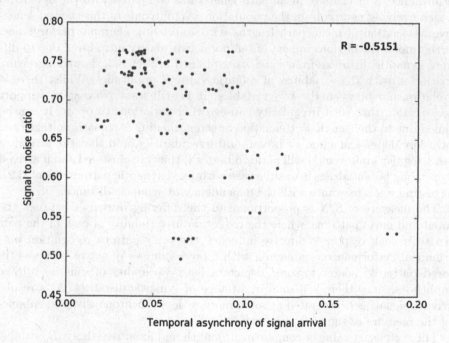

Fig. 8. Temporal irregularity of signal arrival at soma and its impact on the S/N ratio.

4 Discussion

The pattern recognition capacity of a cell was assessed in an artificial neural network model of an associative memory, discerning between two sets of transient input patterns. To estimate the likely effect of different dendritic topologies and the impact of spatio-temporal noise on neuronal performance, a compartmental model simulated the postsynaptic integration of synaptic inputs. The performance of a neuron is directly associated with the spatial distribution of its synapses. Consequently, it performs sufficiently well when inputs are clustered as opposed to when they are purely random. Persistent and clustered synaptic inputs to a specific region of a dendrite enhances the strength and arrival timing of these signals at soma, complying with the Hebbian learning rule for long-term potentiation and synaptic plasticity. However, when inputs are randomly generated, a neuron receives these signals widely across its dendritic arbour, attaining apportioned synaptic weights to exhibit its average discriminatory performance. The EA successfully generated best possible individuals in each generations and the evolved neurons performed sufficiently well compared to the originally generated neurons. Considering the impact of different dendritic distance on neuronal performance, a measure of mean path length was compared with the S/N ratio of each evolving neuron from the population. An outcome of this showed a negative association of mean path lengths with improving neuronal performance. Furthermore, the obvious impact of temporal irregularity introduced due to different synaptic distances from soma, as explained in Sect. 3.3, showed a growing association with the variability of somatic voltage amplitudes. Whilst there is a relationship between the two variables, it is still least relevant to support the assertion that such irregularity causes other variations to occur. It can be argued due to the fact that attenuation or strengthening of somatic voltage does not solely dependent upon such temporal irregularities, and that the arrival of such irregular inputs could still perform better if they are clustered and if arrived with strong potentials, as it was the case with fixed synaptic patterns (Sect. 3.2), suggesting weak relevance with the dependency of synaptic distance.

The measure of S/N is proportional to the differing responses between the stored and novel patterns, where the reduction in variability of each of the pattern set's peak responses directly influence the cell's pattern recognition performance. Performance of a neuron with a large number of active synapses (in stored patterns) nearer to soma depicts a least variability of somatic voltage amplitudes, since there is a smaller variance of synaptic distances. As a result, arriving signals get integrated at soma more or less synchronously, also enhancing the measure of signal to noise ratio.

The increasing value of compartment length and axial resistivity, R_a exhibit a degrading and prolonged amplitude of somatic voltage. To verify these effects, different values such as 200, 300, 400 μm for compartment length and 50, 100, 150 Ωcm for R_a were considered, which showed decreasing mean stored (μ_s) responses; −57.02, −59.79, −61.20 (each affected by an increasing compartment length) and −57.70, −59.35, −60.31 (each affected by an increasing R_a) respectively. An appropriate value of axial resistivity is quite uncertain and it was estimated between 50–400 Ωcm [4]. A relatively lower value of R_a improves the S/N ratio by decreasing intracellular resistance for ions to move, ultimately reducing the variability of somatic responses to stored and novel patterns.

The present study is an extension of work done by Graham [4] and De Sousa [3] in which synaptic patterns were composed of synchronously arriving randomly generated transient inputs. A study carried out by Graham [4] involved pattern recognition of 100 and 200 active inputs, which in the current study are considered as stored and novel patterns, but with equal number of active inputs. Currently, a notion of clustered synaptic inputs is used in addition to the randomly generated input patterns along with an EA to optimise number of inputs that each active synapse receives. Clustered synapses aim to minimise the distance variations of inputs to soma and grow the membrane voltage in each cluster by controlling the strength of combined signals - which as a result produce least variable EPSP amplitudes for an improved S/N ratio. The implication of this findings is that the spatial organisation of active inputs determines dendritic integration of a postsynaptic neuron with a little dependency on any specific structure of a morphology. In other words, a correlated and synchronous arrival of inputs to soma is the best predictor for high performing neurons.

Acknowledgements. I would like to express my sincere gratitude to Dr. Rene te Boekhorst for his valued support and guidance extended to me.

Appendix

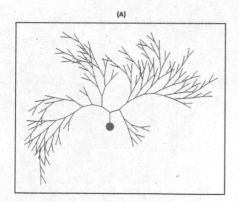 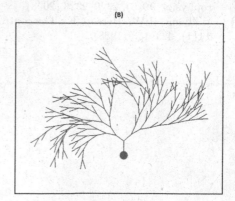

Fig. 9. Dendritic morphologies evolved after 100 evolutionary iterations to recognise random and clustered input patterns in (A) & (B) respectively.

References

1. Cooke, S.F., Bliss, T.V.P.: Plasticity in the human central nervous system. Brain **129**(7), 1659–1673 (2006)
2. Cuntz, H., Borst, A., Segev, I.: Optimization principles of dendritic structure. Theoret. Biol. Med. Model. **4**(1), 1 (2007)
3. De Sousa, G., et al.: Dendritic morphology predicts pattern recognition performance in multi-compartmental model neurons with and without active conductances. J. Comput. Neurosci. **38**(2), 221–234 (2015)
4. Graham, B.P.: Pattern recognition in a compartmental model of a CA1 pyramidal neuron. Network Comput. Neural Syst. **12**(4), 473–492 (2001)
5. Gulledge, A.T., Kampa, B.M., Stuart, G.J.: Synaptic integration in dendritic trees. J. Neurobiol. **64**(1), 75–90 (2005)
6. Hebb, D.O.: The Organization of Behavior: A Neuropsychological Theory. Psychology Press, New York (2005)
7. Hines, M.L., Nicholas, T.: The NEURON simulation environment. Neural Comput. **9**(6), 1179–1209 (1997)
8. Ho, V.M., Lee, J.-A., Martin, K.C.: The cell biology of synaptic plasticity. Science **334**(6056), 623–628 (2011)
9. Langton, C.G.: Artificial Life: An Overview. MIT Press, Cambridge (1997)
10. Martínez-Cerdeño, V.: Dendrite and spine modifications in autism and related neurodevelopmental disorders in patients and animal models. Developmental Neurobiology (2016)
11. Takeuchi, T., Duszkiewicz, A.J., Morris, R.G.: The synaptic plasticity and memory hypothesis: encoding, storage and persistence. Phil. Trans. R. Soc. B **369**(1633), 20130288 (2014)
12. Van Pelt, J., Verwer, R.W.H.: Growth models (including terminal and segmental branching) for topological binary trees. Bull. Math. Biol. **47**(3), 323–336 (1985)
13. Wen, Q., Chklovskii, D.B.: A cost-benefit analysis of neuronal morphology. J. Neurophysiol. **99**(5), 2320–2328 (2008)
14. Williams, R.W., Herrup, K.: The control of neuron number. Annu. Rev. Neurosci. **11**(1), 423–453 (1988)

Semantic Clustering for Identifying Overlapping Biological Communities

Hassan Mahmoud[1,2(✉)], Francesco Masulli[1,3], and Stefano Rovetta[1]

[1] DIBRIS - Dip. di Informatica, Bioingegneria, Robotica e Ingegneria dei Sistemi
Università di Genova, Via Dodecaneso 35, 16146 Genova, Italy
{hassan.mahmoud,francesco.masulli,stefano.rovetta}@unige.it
[2] Faculty of Computer Science, Benha University, Benha, Egypt
hassan.mahmoud@fci.bu.edu.eg
[3] Sbarro Institute for Cancer Research and Molecular Medicine College of Science
and Technology, Temple University, Philadelphia, PA, USA

Abstract. Proteins encoded by the genes associated with a common disorder interact together, participate in similar pathways, and share Gene Ontology (GO) terms. Drug discovery for certain disease may arise from a hypothesis that genes contributing to a common disorder have an increased tendency for their products to be linked at various functional levels. This may be induced from experimental studies of protein-protein interactions, co-regulation, co-expression, and annotated semantic information (e.g., those stored in Gene Ontology). Our aim is to improve the quality of aggregation discovery in dense biological interactions by incorporating such information embedded in biological repositories and mapping them in the spectral embedding space.

Keywords: Biological community detection · Semantic enrichment · Clustering ensemble · Fuzzy clustering · Spectral clustering

1 Introduction

Protein-protein interactions (PPIs) occur when two or more proteins bind together in a cell in vitro or in a living organism. The interaction interface of proteins is evolved to a specific purpose. Interactions between proteins are important for the majority of biological functions. Not all possible PPIs will occur in any cell at a given time. Proteins involved in the same cellular processes often interact with each other. Therefore, the functions of uncharacterized proteins can be predicted through comparison with the interactions of similar known proteins, and the detection of pertinent communities in PPIs networks can be used to predict the function of uncharacterized proteins based on the functions of others they are grouped with.

In many studies of biological networks, such as *Saccharomyces (S.) cerevisiae* PPIs [11] networks we analyze in this paper, community detection techniques are used to extract aggregations showing dense relationships. The most used community detection algorithms can be categorized into graph based partitioning

© Springer International Publishing AG 2017
A. Bracciali et al. (Eds.): CIBB 2016, LNBI 10477, pp. 235–247, 2017.
DOI: 10.1007/978-3-319-67834-4_19

(bisection), hierarchical clustering, partitional clustering [7], spectral clustering [17], edge removal, and modularity based methods [21].

We propose here a method for community detection based on spectral and fuzzy clustering and exploiting semantic information able to infer the possible overlaps between protein communities in networks, and we study its application to the analysis of the *S. cerevisiae PPI* network.

This paper is organized as follows: Semantic and spectral modularity graph based clustering approaches for community detection are presented in Sects. 2 and 3; in Sect. 4 we describe the semantic clustering based Fuzzy C-Means Spectral Modularity community detection method, while its application to the discovery of communities in the PPI network of *S. cerevisiae* is shown in Sect. 5; Sect. 6 contains the conclusions.

2   Measuring Similarity Based on Semantic Clustering of Semantic Terms

Nowadays, several biological repositories including protein data banks such as *KEGG*[1], or Gene Ontology (*GO*) [1] contain annotated information and biological knowledge either extracted with the help of experts or from papers or other sources. This annotated information can be used to improve the effectiveness of community detection in biological networks using similarity measures entailed in those frameworks.

Lord et al. [13] demonstrated the feasibility of using semantic similarity measures in a biological setting. In this study, the *GO* (semantic) similarity between two proteins was calculated as the semantic similarity of their annotated *GO* terms. They also noticed a strong correlation between protein *GO* similarity using annotations from the UniProt/SwissProt database[2] and their sequence similarity.

The semantic similarity measures between two terms a and b can be categorized into:

– **Feature-based** [24] obtained by extracting the shared features between terms and estimating the weighting parameters that balance the contribution of each feature (often not available)
– **Edge-based** [23] that depend on the path length linking the terms in an ontology and then showing low discriminant power for detailed or wide ontologies;
– **Information-Content-based** (or **Node-based**) that rely on estimating the semantic similarity between two terms or nodes a and b based on the amount of mutual information they share.

In their turn, the information content based approaches can be classified into *annotation-based* and *topology-based*. They evaluate the similarity between two terms or nodes a and b based on the amount of mutual information they share, such as:

---

[1] http://www.genome.jp/kegg/.
[2] http://www.uniprot.org/.

– Resnik's similarity measure [26] that estimates the information content (IC) of Least informative Common Ancestor (LCA), where LCA is a term in the Ontology having the shortest distance from the two terms being compared:

$$sim_{res}(a,b) = IC(LCA(a,b)), \tag{1}$$

where $IC(a) = -log(P(a))$. However, Resnik's similarity may result in inconsistent similarities as it ignores significant graph characteristics and considers only the information content of LCA.

– Lin's similarity measure [12] that enhances Resnik's similarity by considering the information content of LCA and the two compared terms:

$$sim_{lin}(a,b) = \frac{2sim_{res}(a,b)}{IC(a) + IC(b)}. \tag{2}$$

– Jiang and Conrath's similarity measure [10] that is similar to Lin's one:

$$dis_{j\&c}(a,b) = IC(a) + IC(b) - (2sim_{res}(a,b)). \tag{3}$$

– The eXtended Graph based Similarity Measure (XGraSM) [5] is a hybrid approach in which the features of both the parent and child terms of GO are taken into account and is one of the most powerful annotation based semantic similarity measures [19]. In XGraSM **all** the Informative Common Ancestors ($ICA(a,b)$) are considered when computing the semantic similarity between two different terms a and b in GO, and the score between a term and itself is set to 1.

XGraSM shows a higher correlation between protein families and semantic similarity on all aspects of GO unlike the aforementioned (LCA) based measures, namely Eqs. 1, 2, 3 [10,12,26] that rely on the Least informative Common Ancestor (LCA) only.

XGraSM derives new semantic similarities such as $sim_{resnikXG}$, sim_{linXG}, and $sim_{j\&cXG}$ by estimating the average information content between $ICA(a,b)$) terms t. Hence the overall information content $IC(a,b)$ as well as $sim_{resnikXG}$ are given by:

$$sim_{resnikXG} = IC(a,b) = \max_{t}\{IC(t)|t \in ICA(a,b)\}. \tag{4}$$

– GO-universal [18] is a topology based information content measure in which the child is expected to be more specific than its parents. The more children a term has, the more specific its children are compared to that term, and the greater the biological difference. In addition, the more parents a term has, the greater the biological difference between this term and each of its parent terms.

Let N_{GO} is the set of GO terms and links, $(a,b) \in L_{GO}$ represents the link or association between a given parent a and its child b, and the level of the link (a, b) is the level of its source node a, $[a, b] \in N_{GO}$ indicates that the level of term a is lower than that of b.

The Topological information IC_T of a given term $z \in N_G O$ is given by $IC_T(z) = -ln(\mu(z))$ where $\mu(z)$ is a topological position characteristic of z, recursively obtained using its parents gathered in the set $p_z = \{a : (a, z) \in L_G O\}$ and it is given by:

$$\mu(z) = \begin{cases} 1 & \text{if z is a root,} \\ \prod_{a \in p_z} \frac{\mu(a)}{c_a} & \text{otherwise,} \end{cases} \tag{5}$$

where c_a is the number of children of parent term a.
GO-universal [18] is given by:

$$sim_{GOu}(a, b) = \frac{IC_T(a, b)}{max\{IC_T(a), IC_T(b)\}}, \tag{6}$$

where $IC_T(a, b) = -ln\mu(a, b)$. $sim_{GOu}(a, b)$ induces a distance $(d_{GOu}(a, b))$ or a metric based on information theory that is given by: $d_{GOu}(a, b) = 1 - sim_{GOu}(a, b)$. The more topological information two concepts share, the smaller their distance and the more similar they are.

3   Fuzzy Spectral Modularity

Network modularity is used for measuring the strength of community structure in networks and also as an objective function to maximize with suitable optimization methods. Q is a scalar value ranging between -1 and 1. Networks with high modularity imply the existence of dense connections within communities and of sparse links between them. Although modularity suffers a resolution limit especially in the case of small communities, it has the advantages of not requiring prior knowledge about the number or sizes of communities, and it is capable of discovering network partitions composed of communities having different sizes. *Network modularity* (Q) [21] is defined as:

$$Q = \frac{1}{2m} \sum_{i,j} \left[A_{ij} - \frac{k_i k_j}{2m} \right] \delta(c_i, c_j) \tag{7}$$

where A_{ij} is the weight of edge linking vertices i and j, $k_i = \sum_j A_{ij}$ is the degree of vertex i, c_i is the community to which node i is assigned, $m = \frac{1}{2} \sum_{ij} A_{ij}$, and $\delta(c_i, c_j)$ function is 1 if c_i is the same as c_j and 0 otherwise.

Spectral clustering refers to methods used to cluster n objects based on the evaluation of the Laplacian matrix obtained from the data similarity matrix (which is symmetric and non negative), and then in application of a clustering technique (such as K-Means) to data in a subspace spanned by the first k eigenvectors of the Laplacian matrix. Several approaches exploit spectral theory for clustering, such as un-normalized spectral clustering by Shi and Malik [7], normalized spectral clustering by Ng et al. [22], random-walk spectral clustering by Melia and Shi [7].

The *Fuzzy C-Means Spectral Clustering Modularity* (FSM) community detection method introduced in [16] applies the following three improvements to the original Ng et al. [22] spectral clustering algorithm, when used to detect communities in networks:

1. First, the estimation of the number of clusters k is performed by the modularity maximization procedure presented by Newman and Girvan in [21]; the estimated number of clusters is applied for both selecting the top eigenvectors of the Laplacian matrix, and setting the number of clusters for the clustering algorithm.

2. Then, clustering in the affinity subspace spanned by the first k eigenvectors is performed with the application of the Fuzzy C-Means (FCM) clustering algorithm [4] instead of K-Means (used in [22]). As FCM considers that an instance may belong to two or more clusters at the same time, with different membership degrees, this choice supports the detection of overlapping communities and can allow us to understand the role that each protein may play in different communities. The distortion that is minimized by the *FCM* is defined as:

$$\sum_l \sum_j (U_{lj})^m \|x_l - y_j\|^2, \tag{8}$$

with m a fuzziness parameter, $m > 1$, again subject to $\sum_j U_{lj} = 1 \; \forall l$.

3. Following FCM, we applied the spreadability measure ξ [15] as a threshold to remove the nodes with low membership. Spreadability refers to the node capability of spreading information among the different communities belonging to a network. A node s has a high ξ if it belongs to more than one community. Such nodes affect the network flow and information broadcasting in different communities. The spreadability measure depends on the dispersion in node memberships.

It is calculated using the following steps:

(a) For each node s, having membership $U_{1..k}(s)$ in k communities and standard deviation σ, we measure the spreadability cut given by:

$$\varpi = \sigma(U_{1..k}(s)) - \sigma^2(U_{1..k}(s)). \tag{9}$$

(b) Assign s to each community c_i having membership $> \varpi$, then estimate the number of belonging communities given by:

$$\lambda_s = |I_{/s}|, \; I_{/s} = |\{c_i | U_{ci}(s) > \varpi\}|. \tag{10}$$

(c) Nodes having $\lambda > 1$ are identified as fuzzy, and the more the $\lambda > 1$ the more the node is spreadable (that is, has significant influence across the network communities), while nodes having $\lambda = 1$ are referred as crisp (that is, their influence does not extend beyond their own communities).

(d) Spreadability for a fuzzy node s belongs to λ overlapping communities is given by:

$$\xi = \sum_{i=1}^{\lambda} U_i(s), \tag{11}$$

s.t, s is member in c_i, while for crisp node is given by:

$$\xi = 1 - max(U_{1..k}(s)). \tag{12}$$

It is worth noting that ϖ provides s a robust and global fuzzy community identification criterion, and unlike other such as mean $\mu = \sum U_{1..k}(s)$, which is sensitive to noise and membership variation. Moreover, it does not have the limitation of other measures, e.g., the method in [27], exponential entropy given by $\chi(s) = \Pi_{i=1}^{k} u_i(s)^{-u_i(s)}$, and the bridgeness score in [20] given by $b(s) = 1 - \sqrt{k\sigma^2(U_{1..k}(s))}$, that requires an external parameter choice for tuning significant memberships [9].

4   The SC-FSM Community Detection Method

The Semantic Clustering Fuzzy Spectral Modularity community detection method (*SC-FSM*) measures the semantic similarity using both annotation and topology basis by employing an ensemble of *XGraSM* and *GO-universal* (see Sect. 2) similarities to characterize the analyzed proteins. We use the evidence accumulation coding to build a consensus similarity exploiting the protein-protein interaction weights known experimentally from [11] and we apply *FSM* to the consensus similarity matrix for detecting the communities in spectral space.

The annotation-based measures use annotations of related semantic terms, while the topology based approaches consider the intrinsic topology of GO. After employing the semantic enrichment of data using the annotation technique based on *XGraSM* method [5] and Resnik similarity, we built a consensus similarity [8] combining this metric with the topological similarity based GO-universal approach [18] as shown in Fig. 1.

As a protein may participate in several biological processes or carry out different molecular functions, it may be annotated to several terms in Gene ontology. To obtain the semantic similarity between two interacting proteins, we can combine semantic enrichment measures using different mixing strategies such as average, maximum, averaging all the best matches, and best match average [2,25].

The Best Match Average (*BMA*) estimates the average of similarity between best matching terms [2]; For two annotated proteins p and q it is the mean of the following two values: average of best matches of *GO* terms annotated to protein p against those annotated to protein q, and average of best matches of GO terms annotated to protein q against those annotated to protein p, it is given by:

$$BMA(p,q) = \frac{1}{2} \left(\frac{1}{n} \sum_{t \in T_p^X} \max_{s \in T_q^X} S(s,t) + \frac{1}{m} \sum_{t \in T_q^X} \max_{s \in T_p^X} S(s,t) \right) \tag{13}$$

where $S(s,t)$ is the semantic similarity score between terms s and t, T_p^X is a set of *GO* terms in X representing the molecular function (MF), biological process

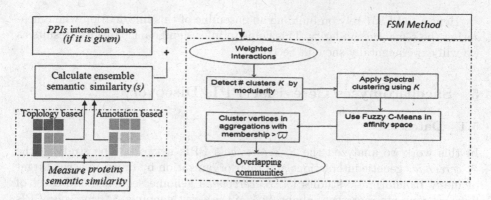

Fig. 1. The proposed semantic clustering fuzzy spectral modularity community detection method (SC-FSM). The method builds an ensemble of the annotation and topology based semantic similarities together with PPI and employ FSM community detection method to infer the overlapping significant k communities with nodes having membership exceeding ϖ.

(BP) or cellular component (CC) ontology annotating a given protein p and $n = |T_p^X|$ and $m = |T_q^X|$ are the number of GO terms in these sets. These two approaches produce different scores and they are equal only when $n = m$, which is not often the case in a set of annotated genes or proteins.

In [19] $sim_{resnikXG}$ together with BMA supported the best results among all the analyzed annotation based approaches, and sim_{GOu} together with BMA mixing strategy showed the best results among all the analyzed topology based approaches We use the same settings in SC-FSM as well as the experiments shown in Sect. 5. The proposed approach could then infer significant interaction communities in the spectral space.

The $XGraSM$ and GO-$universal$ semantic similarity measures were obtained using Proteins interactions and ontology[3], and IT-GOM: Integrated Tool for IC-based GO Semantic Similarity Measures[4].

Note that many studies showed that while analyzing the correlation between semantic similarity and other biological aspects (or dimensions) such as protein pathways, protein complex, protein families ($Pfam$), and gene expression is much more interesting to characterize the semantic relations than sequence similarity in many biological scenarios. Moreover, the information obtained from these aspects may be not be uniform (i.e., in some biological cases some dimensions may contain more interesting information than others), therefore relying on one dimension only may not be sufficient to infer the intrinsic relations between the biological entities (proteins) existing in the network.

---

[3] http://www.lasige.di.fc.ul.pt/webtools/proteinon/.
[4] http://www.cbio.uct.ac.za/ITGOM/tools/itgom.php.

Hence, *SC-FSM* rely on building an ensemble of this information when measuring the semantic similarity. This helps to infer significant biological results as we will experimentally show in Sect. 5.

5   Saccharomyces Cerevisiae PPIs Discovery

5.1   Dataset

In this work we analyzed the *S. cerevisiae*'s PPIs network. The study of the *S. cerevisiae* genetic interactions and their organization by function is the target of many bioinformatic studies [3]. *S. cerevisiae* genome sequence and a set of deletion mutants represents about 90% of the yeast genome. *S. cerevisiae* PPIs can be used to infer regulation of eukaryotic cells. With some 12 million base pairs and 6,466 genes, at least 31% of *S. cerevisiae* genes have a human homologue [3].

We use the *S. cerevisiae* proteins dataset of Krogan et al [11]. In that paper the authors used a tandem affinity purification to process 4,562 different tagged proteins of the yeast Saccharomyces cerevisiae. Each preparation was analyzed by both matrix-assisted laser desorption/ionization time of flight mass spectrometry and liquid chromatography tandem mass spectrometry. Then an ensemble of decision trees was applied to integrate the mass spectrometry scores and to assign the probabilities of protein-protein interactions that were collected in the dataset.

This dataset is an undirected, weighted graph $G = (V, E)$ with V vertices, corresponding to proteins, and E edges indicating protein-protein interaction probabilities (weights) obtained from experiments shown in [11].

We performed our experiments on subgraph from *S. cerevisiae* benchmark dataset having characteristics of 80 proteins and 76 interactions chosen on the basis of prior knowledge about protein involved in different biological process. For instance, protein YAL001C is the largest of six subunits of the RNA polymerase III transcription initiation factor complex (TFIIIC); part of the TauB domain of TFIIIC that binds DNA at the BoxB promoter sites of tRNA and similar genes cooperates with Tfc6p in DNA binding [6].

5.2   Experimental Results and Discussion

We note that many community detection methods such as Newman's edge *betweenness community detection method* [21] are not efficient on large subgraphs because in these situations the random null model underlying modularity becomes unreasonable.

To apply the *SC-FSM* community detection method proposed in Sect. 4, we evaluated the number of clusters k using Newman & Girvan's modularity approach [21] on the analyzed subgraph.

Then we built the ensemble information content semantic similarity by combining both XGraSM [5] annotations, and applying GO-Universal [18] which consider the topological structure in gene ontology. Using the Evidence Accumulation Coding *(EAC)* [8], we built a consensus similarity matrix using the semantic information and *PPI* interaction measurements of [11] study.

Using the proposed *SC-FSM* approach, we could characterize the fuzzy communities by calculating their fuzzy memberships. We consider a protein a belonging to community c if it has a significant membership value, as illustrated in Sect. 4 depicted in Fig. 2.

ID	Protein	1	2	3	4	5
1	YAL001C					
2	YAL007C					
3	YAL011W					
4	YAL016W					
5	YAL017W					
6	YAL019W					
7	YAL021C					
8	YAL026C					
9	YAR002C-A					
10	YAR010C					
11	YBL002W					
12	YBL003C					
13	YBL058W					
14	YBL075C					
15	YBR009C					
16	YBR010W					
17	YBR023C					
18	YBR123C					
19	YBR185C					

ID	Protein	1	2	3	4	5
20	YDL014W					
21	YDL080C					
22	YDL134C					
23	YDL156W					
24	YDL188C					
25	YDR086C					
26	YDR155C					
27	YDR174W					
28	YDR190C					
29	YDR227W					
30	YDR334W					
31	YDR362C					
32	YDR381W					
33	YDR485C					
34	YDR519W					
35	YEL026W					
36	YGL008C					
37	YGL049C					
38	YGL055W					

ID	Protein	1	2	3	4	5
39	YGL190C					
40	YGL200C					
41	YGR047C					
42	YGR161C					
43	YGR231C					
44	YHR132W-A					
45	YIL035C					
46	YIL130W					
47	YJL034W					
48	YJL081C					
49	YKR071C					
50	YKR072C					
51	YLR075W					
52	YLR085C					
53	YLR150W					
54	YLR342W					
55	YML012W					
56	YML041C					
57	YML104C					

ID	Protein	1	2	3	4	5
58	YML109W					
59	YMR072W					
60	YMR214W					
61	YMR273C					
62	YNL055C					
63	YNL121C					
64	YNL209W					
65	YOL012C					
66	YOR014W					
67	YOR058C					
68	YOR061W					
69	YOR110W					
70	YPL007C					
71	YPL128C					
72	YPL146C					
73	YPL152W					
74	YPL216W					
75	YPL224C					
76	YPL235W					

Fig. 2. Fuzzy membership heatmap of the analyzed *S. cerevisiae* proteins in five communities.

In [14] we showed that *FSM* community detection method employed in *SC-FSM* (see Fig. 1) outperforms the state of the art methods in terms of stability of the detected communities, performance, and accuracy using different benchmark networks.

In Fig. 3, we depict the semantic similarity between proteins $(sim_{(a,b)})$ for each of the 80 interactions (edges) in the analyzed *S. cerevisiae*'s PPIs network. For each interaction we measured *XGRASm-Resnik* annotation based similarity, and *GO-universal* topology-based semantic similarity through biological process (BP), molecular function (MF), and cellular component (CC) directed acyclic graphs of gene ontology. The results demonstrate higher correlation between measures in biological process specially because it is better defined than others in the gene ontology.

Moreover, we highlight that the proposed *SC-FSM* extends the depth of community analysis and infers the core interactions in the detected communities those having strong correlation throughout the different biological spaces (aka, dimensions) of gene ontology. For instance, proteins *YAL001C* and *YGR047C* are strongly correlated together in terms of *BP*, *MF*, *CC* dimensions either in the topology space or the annotation space as depicted in Fig. 3.

Idx	From	To	PPI
1	YAL001C	YBL058W	.12
2	YAL001C	YBR123C	.99
3	YAL001C	YDL014W	.10
4	YAL001C	YDR155C	.12
5	YAL001C	YDR362C	.99
6	YAL001C	YDR381W	.30
7	YAL001C	YEL026W	.17
8	YAL001C	YGR047C	.99
9	YAL001C	YML104C	.15
10	YAL001C	YOR110W	.99
11	YAL001C	YPL007C	.15
12	YAL007C	YAR002C-A	.73
13	YAL007C	YBL002W	.25
14	YAL007C	YBL003C	.58
15	YAL007C	YBL075C	.25
16	YAL007C	YBR009C	.72
17	YAL007C	YBR010W	.25
18	YAL007C	YDR086C	.32
19	YAL007C	YDR174W	.72
20	YAL007C	YDR381W	.51
21	YAL007C	YDR519W	.26
22	YAL007C	YGL008C	.83
23	YAL007C	YGL049C	.24
24	YAL007C	YGL055W	.80
25	YAL007C	YGL200C	.77
26	YAL007C	YGR231C	.32
27	YAL007C	YIL035C	.25
28	YAL007C	YIL034W	.83
29	YAL007C	YKR072C	.23
30	YAL007C	YLR075W	.53
31	YAL007C	YLR150W	.33
32	YAL007C	YLR342W	.33
33	YAL007C	YML012W	.93
34	YAL007C	YMR072W	.25
35	YAL007C	YMR214W	.74
36	YAL007C	YNL055C	.33
37	YAL007C	YNL121C	.25
38	YAL007C	YOL012C	.24
39	YAL007C	YOR058C	.23
40	YAL007C	YPL128C	.25

Idx	From	To	PPI
41	YAL007C	YPL146C	.26
42	YAL011W	YBL003C	.63
43	YAL011W	YBR010W	.15
44	YAL011W	YDR174W	.15
45	YAL011W	YDR190C	.99
46	YAL011W	YDR227W	.19
47	YAL011W	YDR334W	.99
48	YAL011W	YDR485C	.99
49	YAL011W	YJL081C	.99
50	YAL011W	YLR085C	.99
51	YAL011W	YML041C	.99
52	YAL011W	YOL012C	.15
53	YAL011W	YPL216W	.19
54	YAL011W	YPL235W	.99
55	YAL016W	YBR023C	.15
56	YAL016W	YBR185C	.17
57	YAL016W	YDL080C	.16
58	YAL016W	YDL134C	.99
59	YAL016W	YDL156W	.16
60	YAL016W	YDL188C	.99
61	YAL016W	YGL190C	.99
62	YAL016W	YGR161C	.55
63	YAL016W	YHR132W-A	.17
64	YAL016W	YIL130W	.11
65	YAL016W	YML109W	.78
66	YAL016W	YMR273C	.93
67	YAL016W	YNL209W	.58
68	YAL016W	YOR014W	.70
69	YAL016W	YPL152W	.36
70	YAL017W	YLR150W	.40
71	YAL019W	YAL026C	.64
72	YAL019W	YDR381W	.15
73	YAL019W	YIL035C	.99
74	YAL019W	YKR071C	.19
75	YAL019W	YML104C	.15
76	YAL019W	YOR061W	.12
77	YAL019W	YPL224C	.15
78	YAL021C	YAR010C	.16
79	YAL021C	YBL003C	.12
80	YAL021C	YBR010W	.12

Fig. 3. Semantic enrichment in yeast S.cerevisiae PPIs of GO biological process (BP), molecular function (MF), and cellular component (CC) aspects. The graphs compare the evaluations obtained using GO-universal and XGraSM.

Figure 4 shows the inferred fuzzy communities obtained in spectral space using *SC-FSM* approach (see Sect. 3). The number of communities ($k = 5$) is obtained using modularity maximization. We depict the fuzzy membership of nodes in Fig. 4 and we adopted the spreadability measure to eliminate the nodes with very low membership to each cluster.

We observe that exploiting annotation and topological structure during semantic enrichment enriched the significance of detected communities.

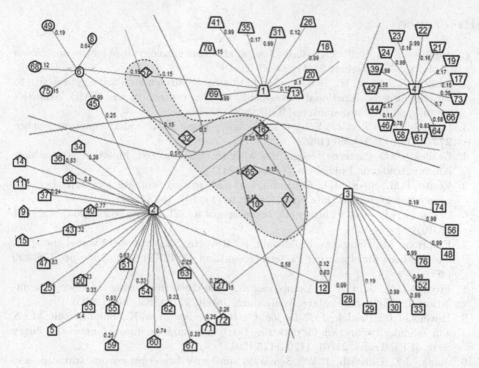

Fig. 4. Results of the *SC-FSM* community detection method on the analyzed *S. cerevisiae* PPIs network. Edges weights are PPIs probabilities. The network is partitioned into five communities. Proteins in the gray region, framed with diamonds, act as bridge nodes with fuzzy memberships.

6 Conclusions

In this paper we proposed a semantics-based fuzzy spectral modularity approach. We performed a semantic enrichment of data using the annotation technique based on the *XGraSM* method [5] and *Resnik's* similarity, and we built a consensus similarity [8] combining this metric with the topological similarity-based *GO-universal* approach [18] with an application to yeast *Saccharomyces cerevisiae* PPIs. The proposed approach inferred five significant overlapping interaction communities in the spectral space.

We conclude that estimating the semantic similarity by combining different sources and biological repositories such as gene ontology or KEGG pathway can boost the detected functional communities. Moreover, mixing the semantic similarity measures together considering different graph characteristics can improve the community detection methods in biological networks.

References

1. Ashburner, M., Ball, C.A., Blake, J.A., et al.: Gene ontology: tool for the unification of biology. Nat. Genet. **25**(1), 25–29 (2000)
2. Azuaje, F., Wang, H., Bodenreider, O.: Ontology-driven similarity approaches to supporting gene functional assessment. In: Proceedings of the ISMB 2005 SIG meeting on Bio-ontologies, pp. 9–10, June 2005
3. Botstein, D., Chervitz, S.A., Cherry, J.M.: Yeast as a model organism. Science **277**(5330), 1259–1260 (1997)
4. Bezdek, J.C.: Pattern Recognition with Fuzzy Objective Function Algorithms. Kluwer Academic Publishers, Norwell (1981)
5. Couto, F.M., Silva, M.J., Coutinho, P.M.: Measuring semantic similarity between gene ontology terms. Data Knowl. Eng. **61**(1), 137–152 (2007)
6. Costanzo, M., et al.: The genetic landscape of a Cell. Science **327**(5964), 425–431 (2010)
7. Filippone, M., Camastra, F., Masulli, F., Rovetta, S.: A survey of kernel and spectral methods for clustering Pattern recognition 41, ISSN: 0031–3203, pp. 176–190 (2008)
8. Fred, A.L.N., Jain, A.K.: Combining multiple clusterings using evidence accumulation. IEEE Trans. Pattern Anal. Mach. Intell. **27**(6), 835–850 (2005)
9. Havens, T.C., Bezdek, J.C., Leckie, C., Ramamohanarao, K., Palaniswami, M.: A soft modularity function for detecting fuzzy communities in social networks. Fuzzy Syst. IEEE Trans. **21**(6), 1170–1175 (2013)
10. Jiang, J.J., Conrath, D.W.: Semantic similarity based on corpus statistics and lexical taxonomy. In: International Conference on Research in Computational Linguistics, ROCLING X, pp. 19–33 (1997)
11. Krogan, N., et al.: Global landscape of protein complexes in the yeast saccharomyces cerevisiae. Nature **440**, 637–643 (2006)
12. Lin, D.: An information-theoretic definition of similarity. In: Shavlik, J. (ed.) Fifteenth International Conference on Machine Learning, ICML Madison, pp. 296–304 (1998)
13. Lord, P.W., Stevens, R.D., Brass, A., Goble, C.A.: Semantic similarity measures as tools for exploring the gene ontology. Pac. Symp. Biocomputing **8**, 601–612 (2003)
14. Mahmoud, H., Masulli, F., Rovetta, S., Abdullatif, A.: Comparison of methods for community detection in networks. In: Villa, A.E.P., Masulli, P., Pons Rivero, A.J. (eds.) ICANN 2016. LNCS, vol. 9887, pp. 216–224. Springer, Cham (2016). doi:10.1007/978-3-319-44781-0_26
15. Mahmoud, H., Masulli, F., Rovetta, S., Russo, G.: Detecting overlapping protein communities in disease networks. In: di Serio, C., Liò, P., Nonis, A., Tagliaferri, R. (eds.) CIBB 2014. LNCS, vol. 8623, pp. 109–120. Springer, Cham (2015). doi:10.1007/978-3-319-24462-4_10
16. Mahmoud, H., Masulli, F., Rovetta, S., Russo, G.: Community detection in protein-protein interaction networks using spectral and graph approaches. In: Formenti, E., Tagliaferri, R., Wit, E. (eds.) CIBB 2013 2013. LNCS, vol. 8452, pp. 62–75. Springer, Cham (2014). doi:10.1007/978-3-319-09042-9_5
17. Mahmoud, H., Masulli, F., Rovetta, S.: A fuzzy clustering segmentation approach for feature-based medical image registration. In: Ninth International Meeting on Computational Intelligence Methods for Bioinformatics and Biostatistics CIBB2012, p. 41 (2012)

18. Mazandu, G.K., Mulder, N.J.: A topology-based metric for measuring term similarity in the gene ontology. Adv. Bioinform. **15**, 195–211 (2012)
19. Mazandu, G.K., Mulder, N.J.: Information content-based gene ontology functional similarity measures: which one to use for a given biological data type? PLoS ONE **9**(12), e113859 (2014)
20. Nepusz, T., Petrczi, A., Ngyessy, L., Bazs, F.: Fuzzy communities and the concept of bridgeness in complex networks. Phys. Rev. E **77**(1), 016107 (2008)
21. Newman, M.E.J.: Modularity and community structure in networks. proc. Nat. Acad. sci. 103(23), pp. 857–869 (2006)
22. Ng, J., Jordan, M.I., Weiss, Y.: On spectral clustering: analysis and an algorithm. In: Proceedings of Neural Information Processing Systems, pp. 849–856 (2002)
23. Rada, R., Mili, H., Bichnell, E., Blettner, M.: Development and application of a metric on semantic nets. IEEE Trans. Syst. Man Cybern. B Cybern. **9**, 17–30 (1989)
24. Rodriguez, M.A., Egenhofer, M.J.: Determining semantic similarity among entity lasses from different ontologies. IEEE Trans. Knowl. Data Eng. **15**, 442–456 (2003)
25. Sevilla, J.L., Segura, V., Podhorski, A., Guruceaga, E., Mato, J.M., Martinez-Cruz, L.A., et al.: Correlation between gene expression and GO semantic similarity. IEEE/ACM Trans. Comput. Biol. Bioinf. **2**(4), 330–338 (2005)
26. Resnik, P.: Using information content to evaluate semantic similarity in a taxonomy. In: Mellish, C.S. (ed.) 14th International Joint Conference on Artificial Intelligence, IJCAI 1, pp. 448–453 (1995)
27. Zhang, S., Wang, R.S., Zhang, X.S.: Identification of overlapping community structure in complex networks using fuzzy c-means clustering. Phys. A **374**(1), 483–490 (2007)

Author Index

Al-Ahmadie, Hikmat A. 42
Alves Coelho, Olívia M. 93
Andrews, Peter W. 93

Batouche, Mohamed 70
Benmounah, Zakaria 70
Bertolazzi, Paola 160
Biga, Veronica 12, 93
Boukelia, Abdelbasset 70
Brdar, Sanja 199
Burzykowski, Tomasz 59

Carrieri, Anna Paola 27
Cazzaniga, Paolo 82
Chen, Weiwei 184
Coca, Daniel 12, 93
Colombo, Riccardo 82, 107
Coulter, Jonathan 118
Crnojević, Vladimir 199
Cumbo, Fabio 82, 160

Damiani, Chiara 82, 107

Felici, Giovanni 160
Fiannaca, Antonino 134
Filippone, Maurizio 145, 184
Fuchs, Thomas J. 42

Gokhale, Paul J. 93

Haiminen, Niina 27
Heffernan, Kevin 209
Hill, Nicholas 184
Hoyle, Andrew 118
Husmeier, Dirk 145, 184

Joseph Sirintrapun, S. 42

Kagdi, Mohammad Ziyad 220

La Paglia, Laura 134
La Rosa, Massimo 134
Liò, Pietro 209

Maati, Bouchera 70
Magalhães-Mota, Gonçalo 172
Mahmoud, Hassan 235
Mason, James E. 93
Masulli, Francesco 235
Mauri, Giancarlo 82, 107
Mendes, Eduardo M.A.M. 93
Messina, Antonio 134

Nekkache, Ikram 70
Nicol, James 118
Niu, Mu 145
Nobile, Marco S. 82
Noè, Umberto 184

Parida, Laxmi 27
Pepe, Daniele 1, 59
Pescini, Dario 107
Pires, Filipa 172

Raposo, Maria 172
Ribeiro, Paulo António 172
Rizzo, Riccardo 134
Rogers, Simon 145
Rovetta, Stefano 235

Schaumberg, Andrew J. 42
Schüffler, Peter J. 42
Scott, Erin 118
Shankland, Carron 118

Teufel, Simone 209

Urso, Alfonso 134

Weitschek, Emanuel 160